Instant Notes *in*

ANIMAL BIOLOGY

The INSTANT NOTES series

Series editor
B.D. Hames
Department of Biochemistry and Molecular Biology, University of Leeds, Leeds, UK

Biochemistry
Animal Biology

Forthcoming titles
Molecular Biology – due Summer 1997
Ecology – due Spring 1998
Microbiology – due Summer 1998
Genetics – due Summer 1998

Instant Notes *in*
ANIMAL BIOLOGY

Richard D. Jurd

Department of Biological Sciences,
University of Essex, Colchester, UK

βIOS
SCIENTIFIC
PUBLISHERS

BIOS Scientific Publishers Ltd
9 Newtec Place, Magdalen Road, Oxford OX4 1RE, UK.
Tel. +44 (0) 1865 726286. Fax +44 (0) 1865 246823
World Wide Web home page: http://www.Bookshop.co.uk/BIOS/

DISTRIBUTORS

Australia and New Zealand
   Blackwell Science Asia
   54 University Street
   Carlton, South Victoria 3053

India
   Viva Books Private Limited
   4325/3 Ansari Road, Daryaganj
   New Delhi 110002

Published in the United States of America, its dependent territories and Canada by Springer-Verlag New York Inc., 175 Fifth Avenue, New York, NY 10010-7858, in association with BIOS Scientific Publishers Ltd.

Published in Hong Kong, Taiwan, Singapore, Thailand, Cambodia, Korea, The Philippines, Indonesia, The People's Republic of China, Brunei, Laos, Malaysia, Macau and Vietnam by Springer-Verlag Singapore Pte Ltd, 1 Tannery Road, Singapore 347719, in association with BIOS Scientific Publishers Ltd.

Production Editor: Priscilla Goldby.
Typeset and illustrated by Florencetype Ltd, Stoodleigh, UK.
Printed by Biddles Ltd, Guildford, UK.

CONTENTS

ABBREVIATIONS

ACTH	adrenocorticotropic hormone	HbA	adult hemoglobin
ADH	antidiuretic hormone	HbF	fetal hemoglobin
ADP	adenosine diphosphate	HRT	hormone replacement therapy
AMDF	anti-Müllerian duct factor	Ig	immunoglobulin
APC	antigen-presenting cell	JGA	juxtaglomerular apparatus
ATP	adenosine triphosphate	LDL	low-density lipoprotein
ATPase	adenosine triphosphatase	LH	luteinizing hormone
cAMP	cyclic adenosine monophosphate	MHC	major histocompatibility complex
CCK	cholecystekinin		
CNS	central nervous system	PIF	prolactin-inhibiting factor
CoA	coenzyme A	PRL	prolactin
CRF	corticotropin-releasing factor	PZ	pancreozymin
CSF	cerebrospinal fluid	Rh	Rhesus factor
DNA	deoxyribonucleic acid	RNA	ribonucleic acid
FSH	follicle-stimulating hormone	T_4	thyroxine
GABA	γ-aminobutyric acid	TCA	tricarboxylic acid
GH	growth hormone	TDF	testicular-determining factor
GIP	gastrin-inhibitory polypeptide	TMAO	trimethylamine oxide
GnRH	gonadotropin-releasing hormone	TRF	thyrotropin-releasing factor
Hb	hemoglobin	TSH	thyroid-stimulating hormone

PREFACE

This book is designed to provide accessible information on animal biology in a compact form for undergraduate students in biology and related life sciences. It is intended to be particularly helpful for revision. The book will be useful for both beginning students and those who are more advanced. In addition, busy lecturers who require a quick reference compendium will find it useful, particularly for tutorial planning.

Instant Notes in Animal Biology is not designed to replace the large comprehensive texts in zoology, comparative physiology or developmental biology which already exist. Nor is it a substitute for lectures, seminars, tutorials or laboratory classes. Rather it is a supplement to all of these, to provide a compendium of core information in a readily accessible form for both ease of learning and rapid revision.

For the student reader, it must be said that no two animal biology courses are the same. Therefore, some of the topics in this book may not be directly relevant to the course being studied. However, scanning through them will certainly help to broaden student insight into the subject. The emphasis of the book is on animal biology. Thus, cell biology, genetics, ecology and animal behavior are not covered here. Readers interested in these subjects are strongly encouraged to read other books in the *Instant Notes* series which address these subjects.

The book is divided into four Sections, each covering a major aspect of animal biology and containing a number of related Topics. Each Topic has main text that describes the subject which is preceded by a list of Key Notes that summarize the main points. The most productive way to use the book is to turn to the Topic of interest and read the main text. Use the Key Notes on that Topic as a memory prompt for revision. Another feature is that each set of Key Notes ends with a list of citations that refer to related Topics. This is an easy way for student readers to navigate the book in a logical way. Finally a list of further reading is provided at the end of the book to guide readers to the literature.

Section A reviews the Animal Kingdom, phylum by phylum. The type of organism is described, followed by the body plan, feeding, locomotion, the skeleton, respiration and the vascular system, osmoregulation and excretion, co-ordination and reproduction, as appropriate, in turn. This is followed by paragraphs on the major groups within the phylum. Examples of animals are given where appropriate. For some phyla there are notes on related, 'minor' animal phyla. A few minor phyla are omitted, and for these students are advised to consult relevant specialist texts. Most biologists now consider that the protozoans belong to their own Kingdom, the Protoctista: I have included them in this book because the heterotrophic protozoans have traditionally been studied with the Animalia.

Zoologists frequently disagree about classification. For example, some consider a particular group as constituting a sub-phylum while others promote it to a phylum or demote it to a class. There is also often disagreement about names (e.g. Monotremata or Prototheria) and even spellings (e.g. Nemertea or Nemertina). The classification used in this book is not necessarily the best one (the only true taxonomic group is the species), and the classifications of others may have equal validity. However, any differences will be quite small and the classification used here is the one I would recommend.

Section B covers a number of co-ordinating principles such as body plans and cavities, skeletal arrangements and symmetry, together with evolutionary issues such as protostomy and deuterostomy, neoteny and pedogenesis, and phylogenetic relationships and origins. Section C describes aspects of comparative physiology, beginning with homeostasis, a unifying concept in physiology. Examples are taken from across the Animal Kingdom, although, for some subjects there is an emphasis on mammals or humans. Functional topics such as locomotion are also reviewed. The topics in Section D review reproductive physiology and developmental biology which here is considered not to end with hatching or birth but to continue into aging.

Simple, yet hopefully clear, Figures and Tables are provided throughout the book. These are not intended to substitute for the more detailed, often multi-colored, illustrations frequently found in large textbooks, but they will be much easier to learn and reproduce. After nearly 30 years of teaching in universities, I am still dismayed by how reluctant some students are to use diagrams, charts and tables in continually assessed work and examinations, yet they are often an easy and very effective way of communicating information. The diagrams included here will give students ideas for producing other diagrams of their own.

The overriding goal of this book, and indeed of the whole *Instant Notes* series, is to present the essential information concerning animal biology in a compact, readily accessible form which lends itself to student learning and revision. If, as a student reader, you use the book for browsing and for revision – and it helps you to understand the subject better and pass those all-too-important examinations and assessments – it will have fulfilled its prime role. Now – read on!

Acknowledgments

I am grateful to many friends at Essex and elsewhere for their forbearance and their helpful discussions and answers to my questions. Particular thanks are due to my colleague and teaching collaborator, Dr Martin Sellens, for his percep-tive and critical advice and also to Professor Christopher Bayne, Oregon State University for reviewing the manuscript. My research student, Michael-Anthony Price, and my elder son, Peter (studying Biology as an undergraduate at Southampton University) have read and commented on a number of chapters from a consumer's viewpoint. I must emphasize, however, that any shortcom-ings remaining in the book are my responsibility: I should be pleased to receive readers' comments. Lastly, I thank Elizabeth, and Peter, Andrew and Mary for their patience and understanding during a year when 'the book' has occupied a large chunk of my life!

Richard D. Jurd

Geological time ladder

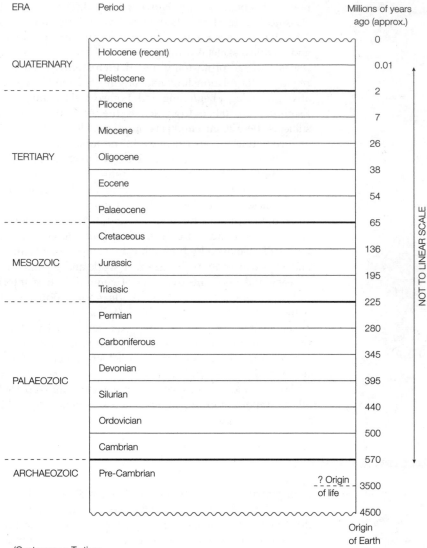

ERA	Period	Millions of years ago (approx.)
		0
QUATERNARY	Holocene (recent)	
		0.01
	Pleistocene	
		2
TERTIARY	Pliocene	
		7
	Miocene	
		26
	Oligocene	
		38
	Eocene	
		54
	Palaeocene	
		65
MESOZOIC	Cretaceous	
		136
	Jurassic	
		195
	Triassic	
		225
PALAEOZOIC	Permian	
		280
	Carboniferous	
		345
	Devonian	
		395
	Silurian	
		440
	Ordovician	
		500
	Cambrian	
		570
ARCHAEOZOIC	Pre-Cambrian	
	? Origin of life	3500
		4500 Origin of Earth

NOT TO LINEAR SCALE

(Quaternary + Tertiary
= Caenozoic period)

A1 THE PROTOZOA

Classification: Kingdom Protoctista (Protista), Protozoa

Key Notes

Features	*Type of organism*: unicellular (or 'noncellular') organisms which belong to several phyla; they probably have a polyphyletic origin. *Feeding*: usually heterotrophic. Food is assimilated from a food vacuole. *Locomotion*: often, but not always motile. *Osmoregulation*: osmoregulate by means of a contractile vacuole. *Reproduction*: usually both asexual and sexual.
Phylum Sarcomastigophora	*Sub-phylum Mastigophora* (with flagella): phytoflagellates are autotrophic, zoöflagellates are heterotrophic. *Sub-phylum Sarcodina* (with pseudopodia): these often possess sophisticated skeletal structures.
Phylum Apicomplexa	Members of this phylum are parasites (mainly sporozoans). They have tubular or filamentous organelles at the apical end of the body.
Phylum Microspora	These are also parasitic, with polar filaments in a spore-like stage.
Phylum Ciliophora	Members are ciliated. They have an outer pellicle with ciliary basal granules and trichocysts and they possess macronuclei (control somatic, nonreproductive functions) and micronuclei (control sexual reproduction).
Related topics	Phylum Porifera (A2) Nonmuscular movement (C28)

Features

Type of organism

Today the Protozoa are usually separated from the Animalia and are placed in the Kingdom Protoctista (Protista), with some algae. However, protozoans possess numerous **animal-like features** and they are therefore included here.

Protozoans are **unicellular** (occasionally described as 'acellular') **eukaryotes**. A few algal protoctistans are multicellular, but the gametes, if they are present, are not produced in gonads, nor do the zygotes develop into embryos.

There are several protozoan phyla, and the classification of the group is rather controversial. Key features in the problematical classification of the protozoans include organelle ultrastructure, locomotory and reproductive strategies, and nucleic acid sequences. At least 60 000 species have been described; sizes range from 2–3 μm to 3 mm, and many protozoans are organizationally very complex, as befits organisms in which one cell performs every function of a living creature.

The ancestral protoctistans were probably amoeboid but lacked many eukaryotic organelles such as mitochondria, chloroplasts and flagella. These may have been acquired later by **differentiation** or by **endosymbiosis** of prokaryotic organisms. Evolution from differing points on a path leading to increased complexity could account for the range of protoctistan/protozoan forms.

Feeding
Most protoctistans are **heterotrophs** in which food is assimilated from a food vacuole.

Locomotion
Many protoctistans are motile, using pseudopodia, cilia or flagella. Many forms are sessile.

Osmoregulation
Osmoregulation in marine forms is usually by means of a contractile vacuole.

Reproduction
Reproduction is usually both asexual and sexual, although some species are known which do not exhibit sexual reproduction (e.g. some amoebas).

Phylum Sarcomastigophora

Sarcomastigophorans comprise approximately 48 000 species; one nucleus is present. Flagella or pseudopodia are used for feeding and/or locomotion.

Sub-phylum Mastigophora
Mastigophorans possess a **flagellum** (= whip) (*Fig. 1*) which beats in one or two planes, with a wave of motion passing from the base to the tip, or vice versa, and generating a reactive, propulsive force. Reproduction is typically by longitudinal binary fission; sexual reproduction is unknown in many species.

   Phytoflagellates (which are also claimed by the plant biologists!) such as *Euglena* spp. often have spindle-shaped bodies covered by a **pellicle.** There is one long, principal flagellum and often a second, short flagellum borne at the anterior end. Chlorophyll in chloroplasts is usually present, and a non-living cell wall is frequently found. *Paranema* has an anterior cell mouth (**cytostome**) and is a heterotroph similar in form to the autotrophic *Euglena*; *Phacus* (*Fig. 1*) lives in the guts of tadpoles. Other autotrophic phytoflagellates include the solitary *Chlamydomonas* and the colonial *Volvox*, and the **dinoflagellates** (which also include heterotrophic forms such as the photoluminescent *Noctiluca*).

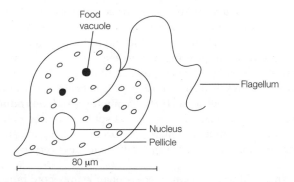

Fig. 1.   Phacus, *a heterotrophic euglenoid phytoflagellate.*

Zoöflagellates lack chlorophyll and are heterotrophic. They include the free-living **choanoflagellates**, with a collar of microvilli around the flagellum base (e.g. the colonial *Codosiga*), and the parasitic **trypanosomatids**. The latter are common parasites of mammals, particularly in tropical countries. They live in the blood and other tissues. The intracellular stage of the life cycle lacks a flagellum, but the extracellular stage has a flagellum extending laterally along the body and anteriorly. The parasites are transmitted by blood-sucking insects. *Trypanosoma brucei* is a cause of African **sleeping sickness** and is transmitted by the tsetse fly. Zoöflagellate hindgut symbionts of termites [e.g. *Barbulanympha* sp. (*Fig. 2*), *Trichonympha* sp.] are among the most complex of protozoans: their cellulases digest wood eaten by their hosts, producing glucose which the termite can then assimilate.

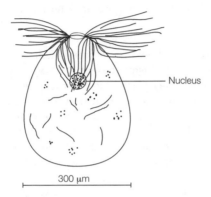

Fig. 2. Barbulanympha, *a zoöflagellate symbiont from termite guts.*

Sub-phylum Sarcodina

The sub-phylum Sarcodina includes **amoebas** and their relatives which possess **pseudopodia**. The pseudopodial condition may be ancestral or may reflect a secondary development from a flagellate stage.

Amoebas may be naked or may be enclosed within a shell which is either secreted or is composed of cemented accretions of minerals, for example *Difflugia* sp. The pseudopodia have a thick outer **ectoplasm** layer which bounds a fluid **endoplasm**. Shape changes reflect changes in the molecular composition of the pseudopodium tip. The pseudopodia surround food particles to form a **food vacuole**; in freshwater species osmoregulation is effected by a **contractile vacuole**.

Some amoebas are commensals or parasites, inhabiting the alimentary tracts of higher animals. *Entamoeba histolytica* (*Fig. 3*) invades human intestinal tissues and causes amoebic dysentry.

Foramenifera, Heliozoa and **Radiolaria** are sarcodine protozoans which possess shells or skeletal structures, often with very elaborate and sophisticated architectures. Long narrow pseudopodia, **axopodia**, are frequently present. The skeletons of these groups are important constituents of some sedimentary rocks.

Asexual reproduction is usually by binary fission. The shells or skeletons of sarcodine protozoans, where present, may be divided or may be given to one of the daughter progeny. Sexual reproduction entails the fusion of two similar

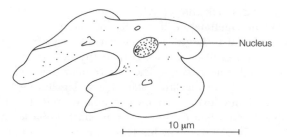

Fig. 3. Entamoeba, *an amoeboid sarcodine.*

gametes. In Foraminifera there may be an alternation of sexual and asexual reproduction stages.

Phylum Apicomplexa

The Apicomplexa is so named because its members have a complex of organelles at the apical end of the organism: these are associated with the protozoans' entry into the cells which they parasitize. Most members of the phylum belong to the taxon formerly known as the **Sporozoa**. Feeding is by **micropores**.

Reproduction usually includes asexual and sexual phases. The infective **sporozoite** invades the host and undergoes asexual fission to form **merozoites**. Merozoites can undergo further multiple fission (**schizogony**), eventually forming gametes which fuse to give a zygote. Meiosis in the zygote results in new sporozoites.

Plasmodium spp. are sporozoans. Four species which parasitize erythrocytes in humans cause malaria: anopheline mosquitos act as intermediate hosts.

Phylum Microspora

These intracellular parasites, common in Arthropoda, lack the apical complex present in sporozoans. There is a spore-like stage, characterized by a polar filament which everts when the spore enters the host. The filament provides a route for the amoeboid sporoplasm to leave the spore. Related phyla include the **Myxozoa** and the **Ascetospora**.

Phylum Ciliophora

The Ciliophora is a large, relatively homogeneous phylum (with about 7000 species); its members all possess **cilia** at some stage in their life histories. The cilia can be used for locomotion and/or feeding. It is speculated that their origins may lie in a group of multiflagellate, multinucleate mastigophorans.

The size of ciliates can extend to 3 mm in length. Most forms are free living in aquatic environments or in the water film microhabitats around soil particles. A few are symbionts (e.g. in the alimentary tracts of vertebrates).

Primitive species can be radially symmetrical (e.g. *Pronodon*), but most are asymmetrical with a distinguishable anterior end. An outer **pellicle** of dense cytoplasm containing the surface organelles maintains the ciliate's shape. Cilia arise from **kinetosomes** beneath the surface of the pellicle. The kinetosomes are connected in long rows by fibrils, the whole assemblage being termed a **kinety**. The surface (somatic) ciliature is primitively found in long rows covering the whole pellicle, but the rows may be reduced to bands, tufts or be absent altogether during some stages in the life cycle. **Trichocysts** in the pellicle can be discharged to transform into threads which can be used for anchorage or for the capture of prey.

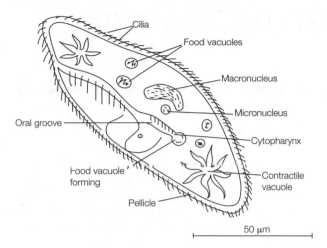

Fig. 4. Paramecium, *a ciliate.*

Locomotion is by synchronized, metachronal waves of ciliary beating down the length of the body, as in *Paramecium* spp. (*Fig. 4*). Some forms, such as *Stentor*, are sessile, the beating of the cilia setting up feeding currents in the water.

Feeding is via a mouth and a **cytopharynx**, leading to the fluid, interior cytoplasm where the food vacuoles are formed. Some ciliates capture their prey directly and can expand their mouths enormously; filter-feeders tend to have more complex buccal apparatus.

Osmoregulation is effected by **contractile vacuoles** at fixed positions within the body. The vacuoles have a ring of radiating tubules which drain water into a central vesicle which rapidly empties, on reaching a certain size, via a pore in the pellicle. Many ciliates can survive drought conditions by **encystment** in which the organism is enclosed in a tough waterproof cyst.

Ciliates have one or more large **macronuclei** which control the non-reproductive, somatic functions of the body. Macronuclei are polyploid and are the main source of ribonucleic acid (RNA). **Micronuclei** are diploid and are involved in reproduction. Ciliates reproduce asexually by transverse fission; the number of asexual divisions seems to be limited before a form of senescence occurs. Organelles may themselves divide, or be resorbed and then re-form.

Sexual reproduction is by **conjugation** involving an exchange and consequent reshuffling of genetic material. Conjugation is necessary to 'rejuvenate' individuals of a clone of asexually produced ciliates, otherwise the organisms are incapable of further asexual division. Conjugation can occur without reproduction, i.e. only the exchange of genetic material takes place. The macronucleus disappears. The micronucleus undergoes two meiotic divisions; all but one of the haploid micronuclei disappear. The remaining micronucleus divides by mitosis to form two 'gametes', one of which migrates to the other conjugating individual to fuse with its stationary micronucleus, thus forming a zygote. The conjugants separate. The zygote nucleus divides by mitosis, one of the products forming the macronucleus.

A2 PHYLUM PORIFERA

Classification: Kingdom Animalia, Sub-Kingdom Parazoa, Phylum Porifera

Key Notes

Features	*Type of organism*: irregular, asymmetrical organisms, with no head, gut or discrete organs; mostly marine but a few freshwater species. *Feeding*: heterotrophic. Water flows through canals and chambers lined with collar cells in chambers, which generates water currents and traps food particles. *Locomotion*: mostly sessile; a few free swimming. *Skeleton*: skeletons of silica (SiO_2) or calcium carbonate ($CaCO_3$) spicules, may or may not contain spongin (protein) fibers. *Respiration*: this is effected by simple diffusion. *Osmoregulation*: this occurs by means of a contractile vacuole. *Reproduction*: most are hermaphrodite but a few sponges have separate sexes; vegetative reproduction is also common. They produce flagellate, free-living larvae.
Classes	Calcarea (calcium carbonate spicules); Demospongiae [spongin (protein) skeleton; silica spicules often present]; Sclerospongiae (spongin fibers with calcium carbonate spicules and often silica spicules too); Hexactinellida (glass sponges with silica spicules).
Related topic	The Protozoa (A1)

Features

Type of organism

There are about 10 000 species of sponges, mostly marine, but about 150 species are freshwater. Sizes range from a few millimeters to over 1 m. Grades of sponge structure are recognized (*Fig. 1*), from simple vase-like **asconoid** sponges, through the folded **syconoid** condition to the **leuconoid** condition with elaborate, ramifying channels and chambers: here the surface area for feeding and substance exchange is vastly increased.

Sponges have an irregular, asymmetrical body architecture; there are **two cell layers** separated by a gelatinous **mesenchyme** containing **amoeboid cells** and **skeletal spicules** and **fibers.** There is a hollow interior cavity (**spongocoel**) connecting to the outside by numerous, small **incurrent pores** (**ostia**) and fewer large **excurrent pores** (**oscula**).

Feeding

The spongocoel is lined with flagellate **choanocytes** which move water through the ostia and along the incurrent canals. Food adheres to the collars of the choanocytes and is passed to the cell body, or is engulfed by amoeboid cells

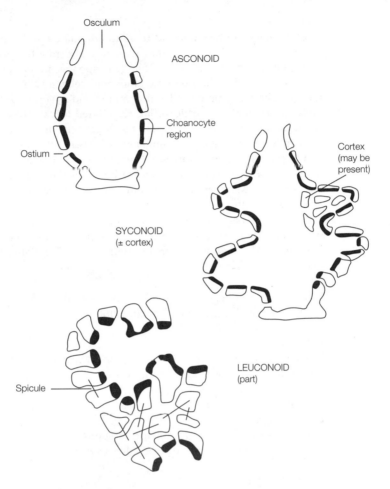

Fig. 1.   Structure of a sponge.

lining the canals. Food products partly digested in the choanocytes may be passed to amoeboid cells, and soluble products diffuse through the sponge.

Locomotion
Almost all sponges are sessile, although a few are free swimming.

Skeleton
Sponges are supported by silica or calcium carbonate spicules, with or without spongin (protein) fibers.

Respiration
Gas exchange takes place through the membranes of the cells by simple diffusion.

Osmoregulation
Freshwater sponges osmoregulate using contractile vacuoles in each cell. Simple diffusion removes other waste substances.

Reproduction
Most sponges are **hermaphrodite** although some species have **separate sexes**. Mesenchymal cells develop into gametes. The motile spermatozoa leave the excurrent opening and enter another sponge individual, are trapped by choanocytes which then become amoeboid cells, and are carried to the ova in the mesenchyme where fertilization occurs and zygotes are formed. The zygotes develop into **motile, multicellular, flagellated larvae**: these metamorphose into the adult form by everting to bring the flagella inside. **Asexual**, vegetative reproduction is seen too: sponge fragments may break off to be transported elsewhere and to grow into new sponges, or **gemmules** consisting of nutrient-laden amoeboid cells (**archaeocytes**) surrounded by epithelial cells may act as 'seeds' for new sponges. Many sponges contain **symbiotic algae** which can be transmitted to progeny in the gemmules.

Sponges are known from early Cambrian deposits, approximately 550 million years B.P.; they are probably a discrete group which is not ancestral to the metazoans, although a relationship with choanoflagellate Sarcomastigophora is likely.

Classes

Four classes of sponges are usually recognized; the Calcarea, the Demospongiae, the Sclerospongiae and the Hexactinellida.

Calcarea
The Calcarea have spicules composed of calcium carbonate (e.g. *Sycon* sp.) The remaining three classes have spicules composed of silica which is extracted from sea water and laid down in special cells.

Demospongiae
The Demospongiae have a skeleton of spongin, a keratin-like protein; siliceous spicules may also be present [e.g. *Microciona* sp.; *Spongilla* sp. (freshwater)].

Sclerospongiae
The Sclerospongiae have spongin fibers, and calcium carbonate (aragonite) spicules in addition (often) to siliceous spicules.

Hexactinellida
These 'glass sponges' have six-rayed siliceous spicules, with the rays developing along three axes mutually perpendicular to each other (e.g. *Euplectella* sp., Venus' flower-basket).

A3 PHYLUM MESOZOA

Classification: Kingdom Animalia, Sub-Kingdom Metazoa, Phylum Mesozoa

---

Key Notes

Features	*Type of organism*: bilaterally symmetrical endoparasites of marine invertebrates; two tissue layers; they lack organs apart from a gonad. *Feeding*: heterotrophic; absorb host fluids osmotrophically. *Locomotion*: ciliated locomotion. *Reproduction*: alternation between asexually and sexually reproducing generations.
Related topics	Phylum Porifera (A2) Phylum Cnidaria (A4)

Features

Type of organism

Mesozoans are **bilaterally symmetrical**, microscopical (0.1–8 mm) 'worms' with **two tissue layers**. Apart from **gonads**, they **lack organs**. An example is *Rhopalura granulosa*, a parasite in clams.

Debatably, in evolutionary terms mesozoans lie between the Protoctista and 'true' Metazoa, although some zoologists consider mesozoans to be degenerate Platyhelminthes. There are two arguably unrelated classes (**Orthonectida** and **Rhombozoa**) which may soon be raised into phyla in their own right.

Feeding

Mesozoans are parasites in marine invertebrates such as cephalopods and polychaetes. They absorb nutrients directly through the body surface (**osmotrophy**) from the urine and other tissue fluids of their hosts.

Locomotion

Locomotion is by cilia.

Reproduction

Mysteries remain concerning the mesozoans' life cycles. Free-living larvae enter their hosts in which they develop into adult forms. There is an **alternation of generations** between sexually reproducing and asexually reproducing forms.

A4 PHYLUM CNIDARIA (= COELENTERATA)

Classification: Kingdom Animalia, Sub-Kingdom Metazoa, Phylum Cnidaria

Key Notes

Features	*Type of organism*: aquatic animals, mostly marine; they may be free-living (medusa) or sessile (polyp). Examples are jellyfishes, sea-anemones and corals. *Body plan*: two cell layers (diploblastic) (see *Fig. 1*) separated by a gelatinous mesoglea. The outer layer of cells is ectoderm or epidermis; the inner layer is endoderm or gastrodermis. They are radially symmetrical with a mouth at one end of the radial axis. They have a single internal cavity, the enteron or gastrocoel, which forms the blind-ending gut (there is no anus). *Movement and locomotion*: muscle cells with fibers attached along their whole lengths to the limiting membrane of the mesoglea. *Feeding*: stinging nematocysts; a gut with no anus, mesenteries increase the surface area in many species. *Co-ordination*: a nerve net is present which co-ordinates swimming of jellyfish and other behavior. *Reproduction*: gonads are aggregations of developing gametes. Fertilization is external, producing free-living planula larvae. Asexual budding is common.
Classes	*Hydrozoa*: these are mainly marine, including colonial forms with considerable polymorphism among polyps, but also including freshwater hydras. *Scyphozoa*: jellyfishes with a dominant medusal stage. *Anthozoa*: sea-anemones, sea-pansies and soft and stony corals; these have a dominant or exclusively polyp form.
Related phylum	*Phylum Ctenophora*: the ctenophorans, or sea-combs, have a diploblastic body plan and eight paddle-like comb-plates.
Related topics	Body plans and body cavities (B1) Skeletons (B3) Symmetry in animals (B2)

Features

Type of organism

Cnidarians, formerly known, together with the Ctenophores (sea-combs), as the **Coelenterata**, include sea-anemones, corals and jellyfishes. They are usually marine although there are a few freshwater species (e.g. *Hydra*). There are two basic forms of cnidarian. The columnar, sac-like **polyps** have an upward-facing mouth around the end of the radial axis, surrounded by tentacles. Most polyps are sessile but some move by gliding (pedal creeping), somersaulting or burrowing into the substratum. **Medusae** resemble umbrellas and are usually pelagic (free swimming), swimming mouth-down with the tentacles floating outwards. Sizes of cnidarians range from the microscopic to 2 m for polyps; the medusae of the sea-blubber *Cyanea* can attain 3.5 m diameter with 30 m tentacles.

Coral reefs, below the surface of shallow seas, are largely made up of the $CaCO_3$ exoskeletons of several species of cnidarians and of other calcium-precipitating organisms such as certain algal species.

The earliest cnidarians recorded are hydrozoans such as *Ediacara* sp., from 700 million years B.P. By the Cambrian period, all modern cnidarian groups were extant, although the phylum is arguably on an evolutionary cul de sac.

Body plan

Cnidaria are **radially symmetrical** animals with two body layers, an outer **epidermis** (ectoderm) and an inner **gastrodermis** (endoderm) which lines the gut, a condition known as **diploblastic** (*Fig. 1*). Between the **gastrodermis** and the outer **epidermis** lies a gelatinous **mesoglea** containing some loose cells. Support is effected by the hydrostatic pressure in the gastrocoel (which acts as a hydrostatic skeleton when the mouth is closed), by the mesoglea and, in some polypoid cnidarians, by a calcareous exoskeleton.

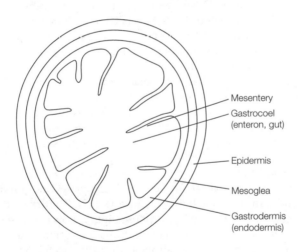

Fig. 1.   Cross-section through a typical cnidarian.

Locomotion

Muscle cells lie in the mesoglea. Sheets of radial and circular muscle fibers effect medusal swimming. Polyps are sessile but can move by burrowing, somersaulting or pedal creeping.

Feeding

The **gastrocoel**, lined with endodermal cells, is the sole body cavity: it serves as the gut. Cnidarians are carnivores: they possess stinging cells, **nematocysts**, on their tentacles. The nematocysts discharge when **undulipodia** on the surfaces of the tentacles are stimulated. Food enters the gut via the mouth. The absorptive surface of the gut is increased by **mesenteric** folds. There is no anus, so waste is discharged through the mouth.

Many cnidarians possess intracellular, photosynthesizing dinoflagellates as **symbionts**; some freshwater hydras have similar symbiotic relationships with chlorophyte algae. The presence of such symbiotic organisms has been shown to enhance the growth of the cnidarians.

Co-ordination
Control and integration is facilitated by a **nerve net** with nerve cells possessing unmyelinated, naked cells with highly branched fibers.

Reproduction
Gonads are aggregations of gametes; fertilization is external. Development is usually via a **planula** larva (see *Fig. 3*). Asexual reproduction by budding is common, particularly in colonial forms.

Classes

Three classes of cnidarian are recognized: the Hydrozoa, the Scyphozoa and the Anthozoa.

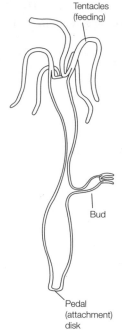

Tentacles (feeding)

Bud

Pedal (attachment) disk

Fig. 2.   A hydrozoan polyp.

Class Hydrozoa
The 3100 species of Hydrozoa are mainly marine and include colonial forms and the fire corals; there are also a number of freshwater **hydras** (*Fig. 2*). Colonial forms (e.g. *Obelia* sp.) may exhibit considerable **polymorphism among the polyps**, with polyps being specialized for functions such as feeding, reproduction and defense. This trend is seen most developed in **siphonophorans** such as *Physalia* sp., the Portuguese man-o'-war, where a **float** (filled with 90% carbon monoxide) aids buoyancy.

Reproduction may be by **asexual budding** of polyps off parent polyps, but **sexual reproduction** is observed too. A medusa forms from a bud on the polyp colony; the medusa floats away and produces free-swimming eggs and sperm. Following fertilization and cell division, a free-swimming, mouthless, ciliated **planula** larva develops which later metamorphoses into a polyp.

Class Scyphozoa
The class Scyphozoa (about 200 spp.) comprises the **jellyfishes** whose dominant life-form is a free-living medusa (*Fig. 3*). The fertilized egg develops into a planula larva which metamorphoses into a polyp: this continually buds over several years to form free-swimming male or female medusae. (Some oceanic scyphozoan planulae metamorphose directly into medusae, thus bypassing the polyp stage.)

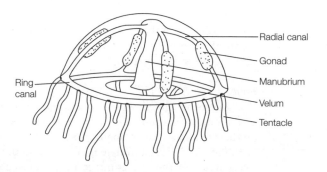

Radial canal

Gonad

Manubrium

Ring canal

Velum

Tentacle

Fig. 3.   A scyphozoan medusa.

Class Anthozoa

The Anthozoa (about 6200 species) are the **sea-anemones, sea-pens, sea-pansies** and the soft and stony **corals**. They are marine animals, solitary or colonial, existing as hermaphrodites or as separate sexes. The stony corals are important constituents of coral reefs.

Anthozoan eggs usually develop into planulae which settle and metamorphose into polyps. No medusae form. Viviparity is observed in some species in which the planula is retained in the body, a mature polyp being released.

Related phylum   *Phylum Ctenophora (sea-combs or comb-jellies)*

There are about 90 species of ctenophorans; these marine, diploblastic animals possess a mesoglea and a branching gastrocoel. Eight paddle-like comb plates (**ctenes**) with aggregates of external cilia give a radial symmetry on which two tentacles superimpose a bilateral symmetry. Stinging **colloblasts** are the equivalent of the cnidarian nematocysts. They possess true muscles but have only a nerve net. Many are phosphorescent, and symbiotic algae within the comb-jellies can give the latter vivid colors. An example of a ctenophore is *Cestum veneris*, Venus' girdle.

A5 PHYLUM PLATYHELMINTHES

Classification: Kingdom Animalia, Sub-Kingdom Metazoa, Phylum Platyhelminthes

---

Key Notes

Features	*Type of organism*: marine, freshwater or terrestrial free-living or parasitic flatworms. Examples are flatworms, flukes and tapeworms. *Body plan*: bilaterally symmetrical, triploblastic animals, greatly flattened dorsoventrally. No enclosed body cavity (acoelomate condition). *Feeding*: incomplete gut (no anus). *Locomotion*: free-living forms use cilia for locomotion. *Skeleton*: hydrostatic skeleton; an outer integument maintains turgor in parasites. *Respiratory and vascular system*: no vascular or other internal transport system. *Osmoregulation and excretion*: by ciliated flame cells and protonephridia. *Co-ordination*: long lateral cords are the main component of the nervous system. Cerebral ganglia are present in free-living worms. *Reproduction*: most are hermaphrodite, exhibiting direct development.
Classes	*Turbellaria*: free-living flatworms which are carnivorous scavengers. *Trematoda*: parasitic flukes with branched guts, often with complex life histories. *Cestoda*: parasitic tapeworms with a head with hooks and suckers, and a body with repeating segments with reproductive organs.
Related topics	Body plans and body cavities (B1) Skeletons (B3) Symmetry in animals (B2)

Features

Type of organism
The Platyhelminthes, the **flatworms and their relatives**, comprise about 15 000 species. Their size varies from less than 1 mm to over 35 m for the longest tapeworms. Marine and freshwater **free-living** platyhelminths are also found as 'terrestrial' forms which live in damp soil (effectively a semi-aquatic micro-habitat utilizing the water films on soil particles), especially in the tropics. Large numbers of species are **parasites** which infest animals in most other phyla.

Body plan
They are **bilaterally symmetrical** animals with a **triploblastic** body plan (*Fig. 1*): the outer **ectoderm** surrounds a **mesoderm** which forms the muscles, reproductive organs and parenchymatous (packing) tissue. This in turn surrounds the **endoderm** which lines the gut. There is **no enclosed body cavity**, a condition known as **acoelomate**.

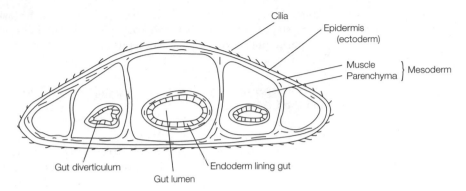

Fig. 1. Cross-section through a typical platyhelminth.

Feeding
The **gut has no anus** (so waste is eliminated via the mouth, as in Cnidaria); the alimentary tract may **ramify extensively** to permit a **large surface area for absorption** of food and distribution of nutrients.

Locomotion
Locomotion in free-living platyhelminths is often by means of **cilia** which are used to glide over a film of secreted mucus. Muscles are also used.

Skeleton
The turgor of the tissues, enclosed in an integument coated with protective glycoproteins in parasitic forms, serves as a hydrostatic skeleton.

Respiratory and vascular system
The **flattened body** gives a **large surface area to volume ratio** which facilitates the uptake of oxygen and the voiding of carbon dioxide and ammonia. There is **no blood vascular system**.

Osmoregulation and excretion
Osmoregulation is by means of **ciliated flame cells** which waft liquid through **protonephridia**, which in turn duct dissolved wastes to the body surface for excretion through pores.

Co-ordination
The **nervous system** is very simple. There is rudimentary **cephalization** (the formation of a head with concentrated nervous and sensory functions) with pigment-cup **eyespots**, a group of **cephalic nerve cells** and **lateral nerve cords**. Tactile and chemoreceptors are associated with the side of the head, sometimes on tentacles or in pits.

Reproduction
Flatworms can **reproduce asexually** by constricting themselves into two parts, each part forming a new worm. They show remarkable capabilities for **regeneration**. **Sexual reproduction** also occurs. Most worms are **hermaphrodite** and, in tapeworms, self-fertilization usually takes place. Yolky eggs are laid in ribboned cocoons; in free-living flatworms they hatch into miniature

adults, but parasitic worms may have complex life histories with successive larval stages and more than one host species in the life cycle. Some flatworms are **parthenogenetic**, with females producing further females without mating with a male.

Classes

Three classes of platyhleminth are recognized: the Turbellaria, the Trematoda and the Cestoda.

Class Turbellaria
The turbellarians are typically small, **free-living 'flatworms'**. They are **carnivorous scavengers** which seize their prey with an **eversible, protrusible pharynx**. The ectoderm is normally **ciliated**. The gut is usually branched, the degree of branching forming the basis for the further sub-classification of this class: **rhabdocoels** have a single-branched gut; **triclads** (which have the widest distribution in terms of habitats) have a three-branched gut; the exclusively marine **polyclads** have a many-branched gut. The Turbellaria may be the ancestral playhelminth group, giving rise to other platyhelminths and perhaps to Nematoda. An example is *Dendrocoelum* sp.

Class Trematoda
The trematodes are **parasitic 'flukes'** (*Fig. 2*) with one or two suckers and often a hooked attachment organ; the gut has two branches. Feeding can be by active transport through surface microvilli or by use of an oral sucker. The ectoderm lacks cilia but may have spines. **Monogenea** have a **simple life cycle** with one vertebrate host and a free-swimming larva; **Digenea** have a **vertebrate host and an intermediate host** (often a mollusc) which harbors the larval stages. Examples are *Opisthorchis sinensis* (Chinese liverfluke), and *Fasciola hepatica* (sheep liverfluke).

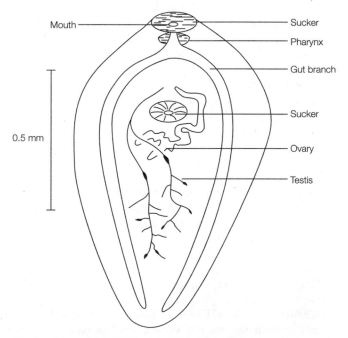

Fig. 2.   A trematode (fluke) platyhelminth, e.g. Fasciola *sp.*

Class Cestoda

The **parasitic tapeworms** (*Fig. 3*) have a head (**scolex**) with hooks and/or suckers, and a body with repeating segments (**proglottids**), each bearing a set of reproductive organs. The mature proglottids detach and so the eggs can be dispersed. The adults are parasites, usually living in the alimentary tracts of vertebrates where they feed by absorption of food through the body surface (**osmotrophy**): the tapeworms lack a gut themselves. There is at least one larval stage which may parasitize invertebrate hosts. An example is *Tænia solium* (pork tapeworm).

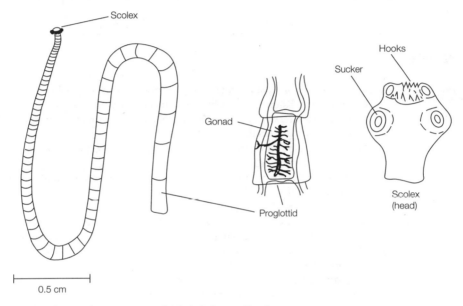

Fig. 3. A cestodan tapeworm platyhelminth, e.g. Taenia *sp.*

A6 PHYLUM NEMERTINA (= NEMERTEA OR RHYNCHOCOELA)

Classification: Kingdom Animalia, Sub-Kingdom Metazoa, Phylum Nemertina

Key Notes

Features	*Type of organism*: mainly burrowing, marine 'worms' (some live in algae or under stones). Examples are ribbonworms and proboscis worms. *Body plan*: bilaterally symmetrical, triploblastic, acoelomate animals; usually dorsoventrally flattened. *Feeding*: complete gut with mouth and anus. The eversible proboscis is sheathed in a rhynchocoel. *Locomotion*: ciliary creeping, or antagonistic contraction of circular and longitudinal muscles. *Skeleton*: hydrostatic, maintained by tissue turgor. *Respiratory and vascular system*: integumentary respiration. The closed vascular system has contractile vessels and blood which frequently contains pigments including hemoglobin. *Osmoregulation and excretion*: flame cells. *Co-ordination*: lateral nerve cords; 'brain'; eyespots. *Reproduction*: usually separate sexes with a larval stage.

Related topics		
	Platyhelminthes (A5)	Skeletons (B3)
	Body plans and body cavities (B1)	Protostomes and deuterostomes (B4)
	Symmetry in animals (B2)	

Features

Type of organism

There are about 900 species of nemertines ranging in length from less than 0.5 mm to more than 30 m. Most ribbonworms are **marine**, living in habitats from the intertidal zone to a depth of 4000 m in the ocean. Some are **tubiculous** (e.g. *Tubulanus* sp. living in a self-secreted mucus tube). There are a few **freshwater** nemertine species and *Geonemertes* sp. lives in **damp soil** in the tropics. Many nemertines exhibit vivid colors and some *Emplectonema* spp. are bioluminescent. Further classification (not covered here) of the phylum is based on the anatomy of the nervous system and the presence or absence of a **proboscis stylet**.

Nemertines are known from **Cambrian** rocks; they may be derived from platyhelminth ancestors. An example is *Lineus longissimus* (bootlace worm).

Body plan

Nemertines are **bilaterally symmetrical, triploblastic acoelomates** (*Fig. 1*). The body is **unsegmented**. There is a basement membrane between the ectoderm

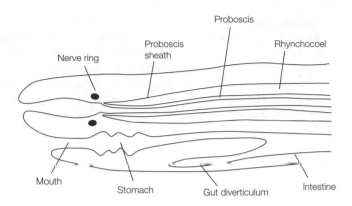

Fig. 1.   Longitudinal section through a nemertine, anterior end, e.g. Lineus *sp.*

and the mesoderm which forms a firm, elastic cover to the **extensive longitudinal and circular muscles**.

Feeding
The nemertines are characterized by a long, anterior, eversible proboscis (used to explore the environment and to 'lasso' prey); when not in use, the **proboscis** is retained within a cavity known as a **rhynchocoel**. Most nemertines are carnivores. Muscular pressure on the fluid in the rhynchocoel everts the proboscis which may have a venomous **stylet** at its tip. Prey are sucked into the mouth, or juices are sucked out. The proboscis may be used for defense and it can be shed, whereupon a new proboscis regenerates. Nemertines have a **complete gut** with a **mouth and an anus.**

Locomotion
Use of **antagonistic longitudinal and circular muscles** facilitates **burrowing**; **swimming** can be effected by lateral undulations of the body; they can **crawl** by **ciliary creeping** using cilia on the integument, or by peristalsis-like waves of the body musculature.

Skeleton
Tissue turgor maintains a hydrostatic skeleton.

Respiratory and vascular system
Nemertines possess a **closed blood vascular system** with **contractile vessels** containing blood which may have **pigments** of different colors: **hemoglobin** is sometimes present. **Respiration** is by **integumentary diffusion**.

Osmoregulation and excretion
Osmoregulation is effected by **ciliated flame cells** lining tubules which lead to surface pores.

Co-ordination
There is considerable **cephalization**. Several hundred **eyespots** may be present. Nemertines have a lobed **'brain'** (cerebral ganglia) and **longitudinal lateral nerve cords.**

Reproduction

Asexual reproduction is by **fragmentation**. **Sexual reproduction** is also seen: the sexes are usually separate. Temporary gonads develop in the parenchymatous mesoderm tissue: these open to the outside by means of gonopores. Eggs are laid in strands in the water where fertilization occurs. The eggs develop (by **spiral cleavage** in a **protostome** fashion, see Topic B4) directly to adults or via a **dipilidium larva**. The hermaphroditic terrestrial *Geonemertes* spp. are viviparous.

A7 PHYLUM NEMATODA

Classification: Kingdom Animalia, Sub-Kingdom Metazoa, Phylum Nematoda

---

Key Notes

Features	*Type of organism*: marine, freshwater, terrestrial or parasitic 'worms' usually with a long, cylindrical body, tapering and rounded at the ends. Adhesive glands or hooks are present. They have complex life histories in parasites. Examples are roundworms, threadworms and hookworms. *Body plan*: bilaterally symmetrical, triploblastic animals. The body is surrounded by a tough elastic cuticle covering an epidermis with longitudinal muscles beneath. They have a pseudocoel body cavity between the longitudinal muscles and the endodermis of the gut. *Feeding*: long, straight gut with a mouth and anus; they have specialized mouth parts, particularly in parasites. *Locomotion*: longitudinal muscles flex the body in anguilliform (eel-like) waves which act against the turgor of the pseudocoel. *Skeleton*: turgor of fluid in the pseudocoel acts as a hydrostatic skeleton. *Respiratory and vascular system*: gas exchange takes place, and vascular systems are usually absent. *Osmoregulation and excretion*: lateral, longitudinal canals open anteriorly. *Co-ordination*: nerve ring around the pharynx and longitudinal nerve cords. *Reproduction*: sexes are usually separate; internal fertilization; eggs or larvae can encyst.
Related phyla	There are a large number of **pseudocoelomate phyla** related to the Nematoda. They are sometimes placed in the 'superphylum' **Aschelminthes**. [Specialist texts review the **Gastrotricha**, the **Nematomorpha** (horsehair worms), the **Acanthocephala**, the **Kinorhyncha**, the **Loricifera** and the **Priapulida**.] The **Rotifera** are discussed in more detail here. In all pseudocoelomates, all muscles are associated with the body wall (ectoderm); the gut (endoderm) lacks muscles. *Phylum Rotifera*: rotifers are free-living animals with a characteristic body design and the presence of a corona of cilia. Reproduction is sexual or by parthenogenesis.
Related topics	Body plans and body cavities (B1) Skeletons (B3) Symmetry in animals (B2) Locomotion: swimming (C30)

Features

Type of organism

There are about 80 000 species of nematode described, although this is probably an underestimate. Sizes range in length from 100 μm to more than 100 cm. Nematodes are usually long, cylindrical 'worms' (*Fig. 1*) with tapered, rounded ends, although some parasitic forms are sac-like in shape. Cilia (except in sense organs) are absent.

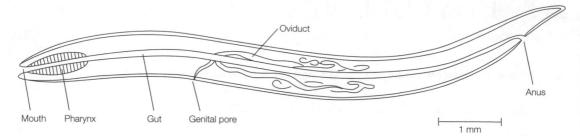

Mouth   Pharynx   Gut   Genital pore   1 mm

Fig. 1.   Longitudinal section through a nematode, e.g. Anguillula *sp.*

Many nematodes are **parasites** in animals and plants. *Trichinella spiralis* causes trichinosis in humans, acquired from eating inadequately cooked pork. Larvae of the nematode encyst in striped (striated) muscle; when this is eaten, the cysts are digested, freeing the larvae into the new host's gut. After about 2 days, the nematodes are sexually mature and males and females mate. The females burrow into the intestinal lining and release enormous numbers of larvae (the females are **ovoviviparous**). The larvae reach the bloodstream having first entered the lymph. From the blood they burrow into muscles and encyst.

Ascaris lumbricoides is a parasite of pigs and humans; the adult lives in the intestine. Eggs are shed in feces and then develop in the soil; re-infection occurs through the ingestion of eggs containing embryos. These hatch in the intestine of the new host and burrow into the wall of the bowel. The larvae enter the lymphatic and then the blood system and pass through the heart to the lungs. From here, they break through to the lung alveoli and thence pass up the bronchioles to the trachea, and then down the esophagus to re-establish themselves in the alimentary tract. **Larvae within the lungs** can lead to **pneumonia; adult nematodes in the bowel** can produce **neurotoxins** and well as causing **physical blockages**.

Free-living nematodes inhabit aquatic and moist terrestrial habitats; they are **important in the ecosystem** for cycling of minerals and food components and for aerating the soil.

Body plan
Nematodes are **bilaterally symmetrical, triploblastic** animals which are covered by a **tough, elastic cuticle**. Segmentation is not present. The **epidermis** surrounds **longitudinal muscles** which in turn surround the body cavity which is a **pseudocoel** (*Fig. 2* shows a general cross-section and *Fig. 3* shows the relationship of muscle fibers to nerve cord; see Topic B1).

Feeding
The pseudocoel surrounds the **long straight gut** which is lined with **endoderm**. The gut may have **specialized mouth parts** such as elaborate, **predatory teeth**, or **hooks** to attach the worm to its host.

Locomotion
The **absence of circular muscles** means that nematodes move by contracting their **longitudinal muscles** to flex the body in **anguilliform** (eel-like) waves.

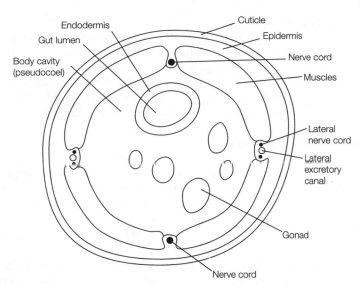

Fig. 2.   Cross-section through a nematode, e.g. Ascaris *sp.*

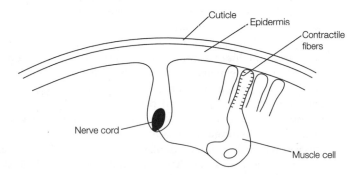

Fig. 3.   The relationship of muscle fibers to nerve cord in a nematode.

Skeleton
The muscles act against the **turgor of the pseudocoel** which serves as a **hydrostatic skeleton** whose constant volume is contained by the thick, tough cuticle.

Respiratory and vascular system
Nematodes do not have respiratory organs: gas exchange occurs through the integument. There is no blood vascular system.

Osmoregulation and excretion
The **excretory system** consists of two **lateral, longitudinal canals** which open by a pore near the anterior end of the worm.

Co-ordination
The **nervous system** comprises a **circumpharyngeal commisure** (a ring around the pharynx) and **longitudinal nerve cords** passing along the body.

Reproduction

The **sexes are separate**, with the male usually being smaller than the female of the species. **Fertilization**, during which the male may grasp the female with copulatory spicules, is **internal.** The **gonads** (with paired or single ovaries or testes) have ducts which exit through a female gonopore, or via the cloaca in the male. Many millions of eggs are produced. These may be very hardy and desiccation resistant. An example is *Ascaris lumbricoides*.

Related phylum   *Phylum Rotifera*

There are about 1500 species of rotifers, most of them living in **fresh water** although a few marine species are known, as well as some which live on damp mosses or in soil; their sizes range from 0.1 mm to 1.0 mm. The body is elongated, with about **1000 cells**, often very constant for each species and/or organ: this phenomenon is called **eutely.** Rotifers have a **trunk** and a **posterior foot**. The anterior end bears a **corona of cilia**, frequently arranged in two disks. The disks beat with **circular waves**, one clockwise and the other counter-clockwise, giving an appearance of wheels. They can **swim** using these cilia, or **crawl** in a hirudiform (leech-like) manner using adhesive glands near the forked foot (which can telescope in many species); a few rotifers are **sessile** and some are **planktonic**, never settling on the substratum. Water currents created by corona may bring in food particles. The pharynx possesses a **mastax** with cuticular teeth projecting into the gut lumen. There is a large stomach and a short intestine leading to the anus.

Reproduction can be sexual using **internal fertilization**. Miniature adults hatch from the limited number of eggs laid (one for each nucleus in the ovary). The males are sexually mature on hatching, the females within a few days. After several generations, **dormant thick-skinned eggs**, that are resistant to cold and desiccation, are produced prior to winter. **Parthenogenesis** ('virgin-birth') is also common and, in some species, no males are known – the parthenogenetic, **thin-skinned eggs** are diploid and produce only females. In some species, when the rotifer population reaches a certain level, haploid eggs are produced and males may hatch. Fertilized eggs are thick skinned and undergo dormancy. An example is *Philodina roseola*.

A8 PHYLUM ANNELIDA

Classification: Kingdom Animalia, Sub-Kingdom Metazoa, Phylum Annelida

Key Notes

Features	*Type of organism*: marine, freshwater or terrestrial segmented 'worms' and leeches, mostly predators or scavengers; they are protostomes. Examples are ragworms, lugworms, fanworms, earthworms and leeches. *Body plan*: bilaterally symmetrical, triploblastic animals whose main body cavity is a coelom formed by a splitting of the mesoderm layer. The body wall has outer circular and inner longitudinal muscles. There is metameric segmentation with segmentally repeating appendages (if present); septa divide the segments. *Feeding*: predators or scavengers, some filter-feeding, tube-dwellers. Straight muscular gut with a mouth and anus. *Locomotion*: antagonism of circular and longitudinal muscles acting against a hydrostatic skeleton; may be aided by lateral processes (parapodia); looping (hirudiform) motion in leeches. *Skeleton*: turgor of coelom provides a hydrostatic skeleton. *Respiratory and vascular system*: gas exchange takes place through the body wall. The closed circulatory system has a dorsal vessel pumping blood anteriorly and a ventral vessel pumping posteriorly. Segmental vessels return blood to the dorsal vessel via the body wall or gut. *Osmoregulation and excretion*: excretion is by segmental nephridia. *Co-ordination*: dorsal anterior cerebral ganglia; circumpharyngeal commisure; paired ventral nerve cord; mainly cephalic sense organs. *Reproduction*: polychaetes with separate sexes with gametes produced in many segments. Protostome development is via a trochophore larva in marine forms; oligochaetes and leeches are hermaphrodite and show direct development.
Classes	*Polychaeta*: mainly marine ragworms, lugworms, fanworms, etc. They possess fleshy, flap-like parapodia with bristles (setae) on each segment. *Oligochaeta*: earthworms and relatives without parapodia; many are terrestrial. *Hirudinea*: leeches with anterior and posterior suckers; many are blood-sucking ectoparasites. They consist of 32 segments. The coelom has botryoidal tissue.
Related phyla	*Echiura*: spoonworms; marine; use a mucus 'net' to trap food. *Pogonophora*: beardworms with a tuft (beard) of tentacles; no gut – food is taken up through the epithelium, or from symbiotic bacteria. *Sipuncula*: burrowing marine worms.
Related topics	Body plans and body cavities (B1) Protostomes and deuterostomes (B4) Symmetry in animals (B2) Sections on physiological functions Skeletons (B3)

Features   *Type of organism*
Annelids are segmented 'worms' that are oval or round in cross-section.
Annelids live in the sea, in brackish and in fresh water and on land, although
they are confined to moist microhabitats. Sizes range from 500 mm to 3 m (the
Australian giant earthworm *Megascolides* sp.). Burrowing annelids are important in the ecosystem because they aerate soil, muds and sands, exposing and
turning over detritus.

Fossil polychaetes are known from the Ediacaran deposits of the Pre-
Cambrian (>700 million years B.P.) and are also well preserved in the middle-
Cambrian Burgess Shales. Polychaetes are probably ancestral to the leeches and
the oligochaetes, and also to the Mollusca.

Body plan
Annelids are **bilaterally symmetrical, triploblastic** animals. An **ectoderm** is
present on the outside; **endoderm** lines the gut lumen. Between the two is **mesoderm**, which forms the muscles, reproductive organs, blood vessels and
excretory organs. The main enclosed body cavity is the **coelom**, formed by the
splitting of the mesoderm (*Fig. 1*). The body is segmented: this repeating,
metameric segmentation is reflected in annular rings seen on the outside of
the body. These delineate the sites of the transverse **septa** which divide one
segment from another.

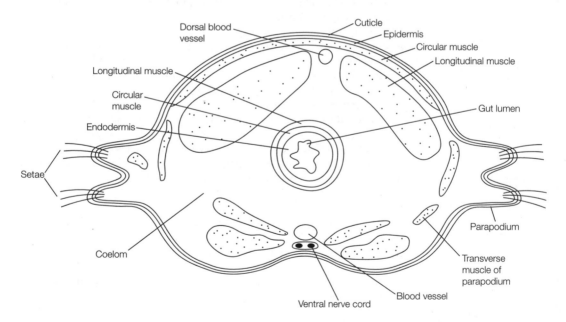

Fig. 1.   Cross-section through a polychaete ragworm, Nereis *sp.*

Feeding
Most are predators or scavengers: swimming forms catch fish larvae. Tubiculous
marine annelids may filter-feed as they lie buried in the substratum: they may
have elaborate fan-like body extensions covered with cilia and mucus for
filtering water containing plankton. There is a straight gut with a mouth and
an anus: food is passed through it by peristaltic contractions of the gut wall
muscles.

Locomotion

The **antagonism of circular and longitudinal muscles** acting against the hydrostatic skeleton facilitates movement; locomotion in polychaetes may be aided by body projections (**parapodia**). Polychaetes and oligochaetes use chaetae (setae) to grip the substratum. Leeches may use a looping, **hirudiform** movement involving their anterior and posterior suckers.

Skeleton

Annelids employ a **hydrostatic skeleton** acting against the turgor of the deformable but constant volume of the coelomic fluid. The segmental division of the coelom facilitates more localized control.

Respiration and vascular system

Gas exchange is through the moist body wall or through extensions of the wall in the form of gills or parapodia. There is a **closed circulatory system** with blood which may contain hemoglobin (erythrocruorin) or chlorocruorin as **respiratory pigments**. The valved vessels contract to propel the blood in one direction, anteriorly in the dorsal vessel and posteriorly in the ventral vessel; segmental vessels return blood to the dorsal vessel via the body wall or the gut wall.

Osmoregulation and excretion

Excretion is via **nephridia**: these ciliated funnels, of which there are two per segment, discharge to the exterior through pores.

Co-ordination

The nervous sytem has **cerebral ganglia** which are linked to a double **ventral nerve cord** by a **circumpharyngeal commisure** (connecting ring). In polychaetes and leeches, **cephalization** is particularly evident: **eyes**, usually with retinas and lenses, are frequent. The whole body may be covered with photosensitive epidermal cells. **Statocysts** associated with balance are situated near the cerebral ganglia.

Reproduction

Polychaetes have separate sexes, with gametes produced in many segments; a **protostome** pattern of development via a **trochophore** larva is seen in marine forms; oligochaetes and leeches are hermaphrodite with direct development

Classes

Three classes are recognized: the Polychaeta, the Oligochaeta and the Hirudinae.

Class Polychaeta

There are about 5400 species of polychaete worms, usually possessing a pair of fleshy, flap-like **parapodia** (*Fig. 1* shows a cross-section of a ragworm) per segment. Each has bristles or **chaetae** (**setae**). Most species are marine, and include **lugworms** (which burrow in sand or mud), sabellid **fanworms** with tubes encrusted with chips of rock and shell, errant **ragworms** which can swim or crawl and burrow in the substratum, and the foreshortened **sea-mice**. Some freshwater and soil species are found. Polychaetes have **separate sexes** and **fertilization is external**. Some adults brood their young: the male may protect and aerate the eggs which may hatch into free-swimming **trochophore larvae**.

Vegetative (asexual) reproduction is seen, whereby individuals are budded off or are transformed into gamete-bearing epitokes. **Regeneration** of lost parts can occur. An example is *Nereis diversicolor* (ragworm).

Class Oligochaeta

The 3100 species of oligochaetes include the **earthworms** and other relatives. Some live in estuarine and fresh waters. They lack parapodia. Cephalization is less marked in this class. Oligochaetes are usually **hermaphrodite**, but each individual mates with another worm. **Asexual budding** and **regeneration** of lost parts is seen, but less so than in polychaetes. An example is *Lumbricus terrestris* (common earthworm).

Class Hirudinea

The 300 species of **leeches**, with anterior and posterior **suckers** (*Fig. 2*), are frequently **blood-sucking ectoparasites** on other animals from many different phyla. There are some free-living predatory and detritivore (detritus-eating) leeches. Leeches lack parapodia and chaetae (setae). The body is usually flattened dorso-ventrally with **32 segments** [although there may be several external grooves (annuli) per segment]. The coelom is restricted by the development of **botryoidal tissue**. Leeches are usually hermaphrodite; one leech will attach to its mate with its suckers and in some species will drive dart-like spermatophores into the partner's body. The eggs develop directly in a desiccation-resistant **cocoon** and hatch as young adults. Leeches do not reproduce asexually nor are they capable of regeneration. An example is *Hirudo medicinalis* (medicinal leech).

Fig. 2. A leech, Hirudo *sp.*

Related phyla

There are three phyla which are related to the annelids: the Echiura, the Pogonophora and the Siphuncula.

Phylum Echiura (spoonworms)

There are about 130 species of echiurans, living on the bottoms of shallow seas where they inhabit rock crevices or burrow into mud and sand; typically they are from 15 to 50 cm long. They are bilaterally symmetrical, triploblastic coelomates. The soft body is covered by a cuticle. Spoonworms feed by using a mucus 'net' to trap food and to sweep it into the ciliated groove at the base of the spoon-shaped, extensible proboscis (which in some species can extend the body's length to about 2 m).

Spoonworms have separate sexes, with extreme sexual dimorphism in some species. Eggs and sperm are released into the coelom where they mature, are collected by a collecting organ and are released into the sea where fertilization occurs. Development follows a protostome manner with spiral cleavage, although segmentation as in Annelida is absent. Examples are *Echiurus* spp.

Phylum Pogonophora (beardworms)

There are about 120 known species of marine, tube-dwelling pogonophorans, a little known group found on the edges of continental shelves. They are bilaterally symmetrical, triploblastic coelomates. They have an anterior tuft or 'beard' of tentacles (*Fig. 3*). The worms live in secreted, chitinous tubes made up of rings. Lengths vary from 10 cm to more than 2 m. Pogonophorans are unusual in lacking a gut: food is taken up through the epithelium or using symbiotic bacteria.

The sexes are separate. Cleavage of the egg is said to be spiral, as in protostomes, although some zoologists find in the Pogonophora affinities with the Hemichordata (which would suggest that they should be considered as deuterostomes). An example is *Siboglinum* sp.

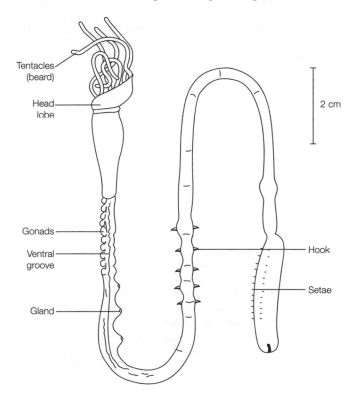

Fig. 3.   A pogonophoran (beardworm), e.g. Siboglinum *sp.*

Phylum Sipuncula (peanut worms)

These are unsegmented, bilaterally symmetrical, triploblastic coelomate, burrowing marine worms: there are about 300 species. Their length varies from about 2 mm to 75 cm. The sexes are separate: sexual reproduction follows a sequence not dissimilar to that described for Echiura (above), except that the gametes are released into the coelom whence they escape to the outside via the excretory **metanephridia** (see Topic C16). An example is *Themiste lageniformis.*

A9 PHYLUM MOLLUSCA

Classification: Kingdom Animalia, Sub-Kingdom Metazoa, Phylum Mollusca

---

Key Notes

| Features |

Type of organism: marine, freshwater or terrestrial, soft-bodied animals, often with a shell. They show a protostome development pattern. Examples are whelks, snails, slugs, mussels, oysters and squids.

Body plan: triploblastic, bilaterally symmetrical, coelomate animals: symmetry may be lost later in development. The body is unsegmented, divided into the head-foot and the visceral mass. A muscular foot is usually involved in locomotion or burrowing. They possess a restricted coelom. Expansion of the open blood system cavities forms the main body cavity or hemocoel. The visceral mass is covered by a mantle extending as a fold(s) enclosing the mantle cavity. The exterior of the mantle usually secretes the shell with one or two valves, sometimes with several plates.

Feeding: complete gut with mouth and anus. A radula is present, also a prominent digestive gland.

Locomotion: a large muscular foot is used (in many species) for creeping and burrowing. Forced expulsion of water from the mantle cavity allows jet-propulsion in cephalopods.

Skeleton: hydrostatic skeleton works against the turgor of the hemocoel; the shell gives support in many species.

Respiration and vascular system: respiration is through the body surface or by gills (ctenidia) in the mantle cavity or through the cavity surface itself. The hemocoel is the main component of the vascular system. There is a dorsal heart.

Osmoregulation and excretion: excretion is by metanephridia which drain the coelom.

Co-ordination: cephalization is usually seen. The nervous system has pedal and visceral cords which join anteriorly to form a cerebral ganglion. Sense organs are frequently present.

Reproduction: separate sexes or hermaphrodite. Gonads are near the pericardial coelom. Trochophore larva develops into a veliger, although in some marine snails the hatching stage is a veliger. Larval stages are bypassed in terrestrial and freshwater species.

| Classes |

Monoplacophora: arguably the ancestral form. They have a single shell, a series of paired gills in the mantle cavity and a radula.

Polyplacophora: chitons with bilateral symmetry and several shell plates.

Gastropoda (Univalvia): one-valved shell (may be lost). This group does include some terrestrial forms (e.g. slugs and snails). The foot is used for pedal creeping.

Bivalvia (Lamellibranchiata): marine species (e.g. oyster), with a few freshwater forms. They show reduced cephalization. Bivalves have a two-valved shell with a hinge, which closes by adductor muscles.

Cephalopoda: advanced marine molluscs (e.g. squid). They possess a radula. The foot is modified to form a funnel and tentacles round the mouth. The shell may be external, internal or absent.

Related topics   Body plans and body cavities (B1)   Protostomes and deuterostomes (B4)
                      Symmetry in animals (B2)   Sections on physiological functions
                      Skeletons (B3)   [e.g. Osmoregulation (C16)]

Features   *Type of organism*
There are about 110 000 species of molluscs, varying in size from less than 1 mm to 1.3 m in the giant clam (*Tridacna* sp.); the giant squid (*Architeuthis* sp.), with tentacles 20 m long, is the largest known invertebrate. They are **soft-bodied animals** which gain support from a **hydrostatic skeleton**, but most have an **internal or external shell**. They possess a **mantle**, a covering of the body wall which lines the shell and secretes the calcium carbonate which constitutes the shell (*Fig. 1*). Most molluscs are aquatic, but some live in damp terrestrial habitats. Molluscs are **protostomes**. A **radula** is characteristic of most molluscs: this hard, chitinous strip bearing teeth is used as a file or a drill to aid food-gathering.

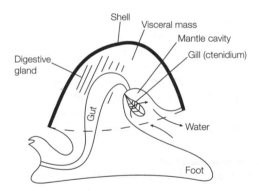

Fig. 1.   A postulated 'primitive' mollusc.

Molluscs' shells are **well preserved as fossils**: monoplacophorans and bivalves are known from the middle Cambrian; univalves and cephalopods from the upper Cambrian. Octopuses are found in the late Cretaceous, but squids are found only in Tertiary rocks. It is thought, on the **basis of their embryology**, that the **Mollusca and the Annelida may share a common ancestor,** although features such as intracellular digestion, external cilia and the architecture of the nervous sytem may point to closer links with the **Platyhelminthes**.

Body plan
Molluscs are **triploblastic, coelomate** animals which are, at least in their early development, **bilaterally symmetrical**. The body is **unsegmented.** The body comprises a **head-foot** and a visceral **mass**. The foot is a large, muscular organ

used, in many forms, for locomotion or burrowing. In the cephalopods it is extended to form arms and tentacles. The visceral mass contains the major internal organs.

The **coelom is much reduced**, but the cavities of the open circulatory system are greatly expanded to form the major enclosed body cavity, termed a **hemocoel** (cf. Arthropoda). In cephalopods, there is no large body cavity as the circulatory system is closed.

Feeding

The **gut is complete**, with a mouth and an anus. Bivalves and some snails are sedentary filter-feeders: the gills are covered by mucus which traps food particles. Cilia waft the mucus to the mouth; many molluscs have a **style** which is rotated by cilia to pull food-laden mucus into the gut. Chitons, slugs and snails use their **toothed radulae** to file at algae and other foodstuffs. Large molluscs such as squids may be active, macrophagous predators.

Locomotion

In many species of molluscs the **foot** is used for creeping or burrowing; cephalopods expel water from the mantle cavity to permit jet-propulsion.

Skeleton

The **hemocoel** is the main component of the molluscan **hydrostatic skeleton**. In many species the shell gives additional support.

Respiration and vascular system

The cavity beneath the shell may contain the **gills (ctenidia)** or, in many terrestrial univalves, is vascularized to form a 'lung'. Respiration is through the body surface (which may be extended to form respiratory tufts), by gills (ctenidia) in the mantle cavity or directly through vascularized mantle cavity surface layers. The hemocoel cavities contain blood in an open circulatory system (in cephalopods the hemocoel is reduced to permit a closed, high-pressure vascular system); there is a dorsal heart.

Osmoregulation and excretion

Excretion is by 'kidneys' or **metanephridia** which drain the coelom.

Co-ordination

Except in bivalves, there is extensive **cephalization,** and **sense organs** include statocysts (for balance), head tentacles, paired eyes and osphradia which act as rheoreceptors monitoring water flow through the mantle cavity. The nervous system comprises paired pedal cords and visceral cords which unite anteriorly to form a **cerebral ganglion** or 'brain'. The cephalopod nervous sytem is arguably the most sophisticated of all invertebrate nervous systems.

Reproduction

In most molluscs, the **sexes are separate** and external fertilization occurs in water; land snails and nudibranch univalves are **hermaphrodite**, although they do not normally self-fertilize. Some oysters can reverse their sex from male to female and back, according to the season. The gonads are found near the pericardial coelom. In primitive molluscs, the gametes exit via the excretory ducts, but more advanced forms have complex gonoducts and other copulatory apparatus. The egg usually hatches to form a **trochophore larva** which metamorphoses into a **veliger larva**, but in many marine snails the egg hatches to form a veliger, while the eggs of cephalopods and land snails bypass the larval stage to develop directly into miniature adults. Some octopuses show parental care of eggs, clams may brood their young and some periwinkle snails are viviparous. Embryonic development follows a **protostome** pattern.

Classes

There are five classes of mollusc: Monoplacophora, Polyplacophora, Gastropoda, Bivalvia and Cephalopoda.

Class Monoplacophora
The little-known monoplacophorans are candidates for being the nearest living relatives to molluscan ancestors. There is a single shell plate with an anteriorly directed, peaked apex. The mantle cavity, containing a series of **paired gills**, forms grooves on either side of the broad foot. There is a radula and a simple digestive system. An example is *Neopilina* sp.

Class Polyplacophora
The 500 species of **chitons** are considered to have diverged early from the ancestral molluscan line. The body retains its **bilateral symmetry**, and the nervous and digestive systems are simple. The **trocophore** larva is similar to that of an annelid. The shell consists of a series of **eight overlapping, anterior–posterior arranged plates**. Chitons are marine herbivores which graze algae from rocks with their rasping radulae; locomotion is by ripples or waves in the muscles of the foot. An example is *Chiton divaceous*.

Class Gastropoda (= Univalvia)
The gastropods (*Fig. 2*) are typically aquatic, but some pulmonate gastropods live on land, mainly in damp places, although desert snails which can estivate during drought periods within their shells are known. The shell, where present, comprises **one piece** or valve. The foot is used for **pedal creeping** using antagonistic muscles creating ripples of contraction. Cephalization is notable, with eyes and tentacles present. There is a radula. Within the shell which is often coiled is a visceral mass. Unisexual and hermaphrodite species exist. The classification of the gastropods is controversial; three orders are usually recognized:

(1) **Prosobranchiata** (e.g. whelks, limpets). These include most gastropods. The majority are marine. The **gills are contained within the mantle cavity in front of the heart**. The shell may be closed by an operculum. The sexes are usually separate. An example is *Nucella lapillus* (dog whelk).

(2) **Opisthobranchiata** (e.g. sea-hares, sea-slugs, sea-butterflies). These marine gastropods have a **reduced or absent shell**. The single gill lies posterior to the heart or may be absent. Most are carnivores living near to the seashore; a few species swim. An example is *Aplysia* sp. (sea-hare).

(3) **Pulmonata** (e.g. snails, slugs). These **freshwater** and **terrestrial** gastropods **lack gills** and respire through a richly vascularized mantle cavity. Shells may be reduced or absent in slugs. They lack an operculum, although a temporary epiphragm may be secreted during a drought. They are usually hermaphrodite herbivores. Examples are *Helix* spp. (land snails).

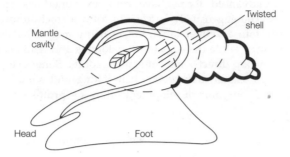

Fig. 2.   A gastropod mollusc, e.g Littorina sp. (periwinkle).

Class Bivalvia (= Lamellibranchiata or Pelycopoda)

Bivalves (*Fig. 3*) are usually marine but there are many freshwater species The **two, hinged valves** of the shell can enclose the body completely. **Cephalization may be very reduced,** with sense organs and the radula being lost. **Gills are enlarged** and possess **ciliary tracts used for filter-feeding**: water is passed over the gills by ciliary action – water enters via the shell gap or through an inhalant syphon; it may leave via an exhalant syphon. The foot can be used for **burrowing** (enzymes to digest lignin and cellulose in wood can, for example, be secreted) or it may be reduced. Powerful **adductor muscles** close the shell. The sexes are usually separate. Examples are *Ostrea* sp. (oyster); *Mya* sp. (clam); *Anodonta* sp. (swan mussel); *Mytilus* sp. (mussel).

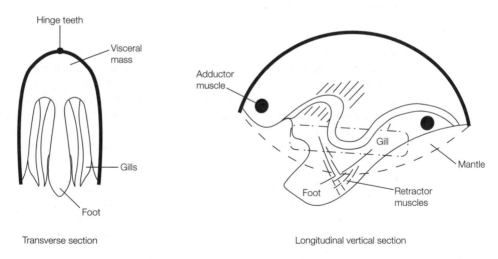

Transverse section

Longitudinal vertical section

Fig. 3.   A bivalve mollusc, e.g Ostrea sp. (oyster).

Class Cephalopoda

These advanced marine molluscs (*Fig. 4*) are active carnivores: they are swim-mers or bottom-livers. The head has a **radula** and there are usually **large, sophisticated eyes** with an architecture similar to that found in vertebrates. The foot is modified to form a funnel and tentacles (with suckers) around the mouth. There may be one or two pairs of gills within the mantle cavity. The **shell** is large and chambered in *Nautilus* sp. and in the extinct ammonites. In most living cephalopods, it is internal (cuttlefishes and squids) or absent (octopuses). A **high-pressure, closed vascular system** with capillaries is present, resulting from a reduced hemocoel: there is **no large body cavity**. Locomotion is by **jet-propulsion** as water drawn into the mantle cavity is forceably expelled via the syphon. Examples include *Nautilus* sp.; *Loligo* sp. (squid).

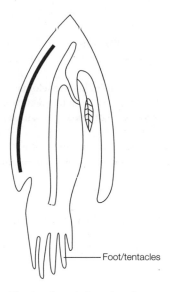

Foot/tentacles

Fig. 4.   A cephalopod mollusc, e.g Loligo *sp. (squid).*

A10 PHYLUM ONYCHOPHORA

Classification: Kingdom Animalia, Sub-Kingdom Metazoa, Phylum Onychophora

---

Key Notes

Features	*Type of organism*: small, caterpillar-like animals which lives beneath stones and in leaf litter. They show protostome development. An example is *Peripatus*.
	Body plan: bilaterally symmetrical, triploblastic coelomate. The body plan is similar to that of the Annelida. They have a reduced coelom.
	Feeding: active predator, possessing a complete gut with a mouth and anus.
	Locomotion: numerous, fleshy limbs.
	Skeleton: the exoskeleton has a cuticle containing chitin.
	Respiratory and vascular system: tracheae, and an open internal transport system.
	Osmoregulation and excretion: nephridia.

Related topics	Phylum Annelida (A8)	Symmetry in animals (B2)
	Phylum Arthropoda (A11)	Skeletons (B3)
	Body plans and body cavities (B1)	Protostomes and deuterostomes (B4)

Features

This small phylum consists of small caterpillar-like animals (*Fig. 1*) which live in leaf litter and beneath stones (the cryptobiotic zone) in the tropics and southern temperate regions. It is tempting to regard them as an evolutionary 'missing link' between the the Annelida and the Arthropoda.

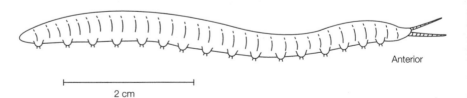

Anterior

|⊢———————————⊣|
2 cm

Fig. 1.   An onychophoran, e.g. Peripatus *sp.*

Onychophorans are bilaterally symmetrical, triploblastic coelomates with a protostome pattern of development. The structure of the **body resembles an annelid** as does the possession of **excretory nephridia**. However, like arthropods, they have a greatly **reduced coelom**, an **open vascular system** and simple **tracheae**. The cuticle contains **chitin**. Numerous short, **fleshy limbs** are seen. Most onychophorans are active predators. **Cephalization** is marked, with prominent **antennae**. An example of an onychophoran is *Peripatus* sp.

A11 PHYLUM ARTHROPODA

Classification: Kingdom Animalia, Sub-Kingdom Metazoa, Phylum Arthropoda

---

Key Notes

Features	*Type of organism*: jointed-legged animals with chitinous exoskeletons. They periodically molt, and show a protostome development pattern. Examples include crustaceans, millipedes, centipedes, insects and arachnids. *Body plan*: bilaterally symmetrical, triploblastic coelomates. They have segmented bodies with two or three distinct parts, and jointed appendages on the segments. The main body cavity is a hemocoel. *Feeding*: many are active predators, but they also include herbivores, fluid-feeders and parasites. They possess a complete gut with a mouth and anus. *Locomotion*: use of jointed limbs for walking. Adult insects use wings for flight. *Skeleton*: chitinous exoskeleton which is periodically molted. Those internal structures which are derived from invaginations of the body wall have a chitinous lining (e.g. foregut, hindgut, tracheae). *Respiration and vascular system*: specialized respiratory organs: gills or tracheae. The vascular system is open with a tubular, dorsal heart. *Osmoregulation and excretion*: discrete excretory organs including Malpighian tubules in insects. *Co-ordination*: the architecture of the nervous system is similar to that of Annelida. *Reproduction*: sexes are usually separate; metamorphosis is common.
Sub-phyla	*Trilobitomorpha*: extinct marine trilobites. They have biramous (two-lobed) limbs. *Crustacea*: mainly marine forms (e.g. crabs). They have biramous appendages, and a skeleton reinforced with calcium salts. The body is usually divided into cephalothorax and abdomen. There are many parasitic forms. Six major classes are noted, including Malacostraca (crabs and lobsters). *Uniramia*: mainly terrestrial. The exoskeleton contains tanned protein. They have uniramous (one-lobed) appendages, walking limbs are behind the head which is sharply set off from the rest of the body. Three classes are noted:

● **Class Diplopoda**: millipedes with two pairs of legs per segment.
● **Class Chilopoda**: centipedes with one pair of legs per segment.
● **Class Insecta**: insects with six walking legs and a distinct head, thorax and abdomen. Apterygota lack wings; pterygota possess wings.

Chelicerata: mostly terrestrial with six pairs of appendages, the first two pairs differing from the other four pairs of limbs; primitively biramous but usually uniramous in land species. The body is divided into prosoma and opisthosoma.

- Class **Merostomata:** marine horseshoe crabs with long, pointed tails.
- Class **Arachnida:** terrestrial scorpions, spiders, harvestmen, ticks and mites.
- Class **Pycnogonida:** sea-spiders with reduced abdomens.

Related phyla

Pentastomida (tongue-worms): endoparasites in the respiratory tracts of carnivorous vertebrates.
Tardigrada (tardigrades): freshwater animals capable of cryptobiosis (suspended animation in the cold).

Related topics

Phylum Annelida (A8)	Skeletons (B3)
Phylum Onychophora (A10)	Protostomes and deuterostomes (B4)
Body plans and body cavities (B1)	Sections on physiological functions
Symmetry in animals (B2)	[e.g. Respiration (C2)]

Features

Type of organism
The Arthropoda (= **'jointed legs'**) is the **largest animal phylum**. Estimates of numbers of species vary from 0.5 to 10 million. Sizes vary from about 0.1 mm to 60 cm long. (Full justice to this phylum, even in outline form, cannot be given in a book of this size.)

Arthropods show a **protostome** developmental pattern. Cilia are not found in Arthropoda.

Body plan
Arthropods are **bilaterally symmetrical, triploblastic, coelomate** invertebrates (*Fig. 1* shows a typical arthropod). Their bodies are usually **segmented**; groups of segments may specialize to form regions such as a head, thorax or abdomen. Each segment primitively possesses a **single pair of jointed appendages** (arguably ≡ polychaete parapodia) used for sensory purposes, feeding, locomotion or reproduction, and may also have a respiratory component.

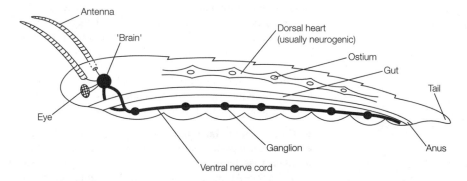

Fig. 1.   Body plan of an arthropod.

In adult arthropods, the embryonic coelom tends to become very reduced, being mainly represented as the cavities in the reproductive organs and in the pericardium. The **main body cavity is a hemocoel.**

Feeding

Many arthropods are predaceous carnivores, but herbivores, fluid-feeders and parasites are also found. There is a complete gut with a mouth and an anus: the foregut and the hindgut are lined with chitin.

Locomotion

The jointed limbs are used for locomotion in water and on land; movement is effected by muscle fibers attached to the inside of the exoskeleton. Insects have wings for powered and/or gliding flight.

Skeleton

Arthropods are enclosed in an **exoskeleton** (organized as an endocuticle, exocuticle and epicuticle) largely comprising a nitrogen-rich polysaccharide, **chitin**, which may be impregnated with calcium salts. The chitin may be rendered more waterproof by waxes (as in insects). Growth is effected by successive **molts** of the exoskeleton.

Respiration and vascular system

Discrete respiratory organs are found in arthropods: gills tend to be used by aquatic species while the more advanced terrestrial species use tracheal systems. The hemocoel provides the blood cavity for the low-pressure, open circulatory system. There is a **dorsal tubular heart** which works on a suction pump principle, and contractions tend to be neurogenic (i.e. nerves stimulate the heart to contract, rather than there being an intrinsic 'myogenic' rhythmicity to the beating of the organ).

Osmoregulation and excretion

Discrete excretory organs are normally found, including green (antennal) glands in Crustacea and Malpighian tubules in Insecta.

Co-ordination

The nervous system resembles that of an annelid, including the possession of a ventral nerve cord.

Reproduction

Arthropod life histories are characterized by many stages, involving molts (ecdysis) of the exoskeleton. There may be profound metamorphosis changes. Sexes are usually separate. Eggs are centrolecithal, with a yolk lying between the central nucleus and the periphery of the egg; cleavage is spiral but commonly superficial.

Sub-phyla   There are four major sub-phyla (or superclasses):

- **Trilobitomorpha**
- **Crustacea**
- **Uniramia**
- **Chelicerata**

although zoologists do not always agree about the details of arthropod classification.

Sub-phylum Trilobitomorpha (trilobites)

Trilobites were a dominant invertebrate form in oceans from the Cambrian to the Carboniferous period. They were oval, flattened arthropods (*Fig.* 2) divided into three longitudinal 'lobes'. All the body segments articulated, allowing the trilobites to roll into a ball. The limbs were biramous (two-lobed), the inner lobe as a walking leg and the outer lobe for swimming, burrowing or perhaps used as a gill.

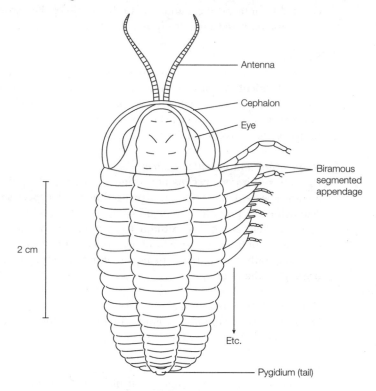

Fig. 2.   A trilobite, e.g. Olenoides sp.

Sub-phylum Crustacea (shrimps, barnacles, copepods, lobsters, crabs, woodlice)

Crustaceans are **mainly marine**, although **freshwater** and a few **terrestrial** species exist. The exoskeleton is reinforced by calcium salts. Appendages are typically **biramous** (two-lobed): there are usually one or two pairs of sensory antennae, followed by biting mandibles. On the segments behind the mandibles lie one or two more feeding appendages, then the locomotory limbs. In the biramous legs, both parts may be used for swimming or the inner part may be used for walking (*Fig.* 3 shows a typical malacostracan arthropod).

The body is usually divided into a **cephalothorax** and an **abdomen**. Respiration is usually by gills, which may be enclosed in a gill chamber (see *Fig.* 4 for a cross-section through a crayfish). Excretion may be by a green gland (antennal gland).

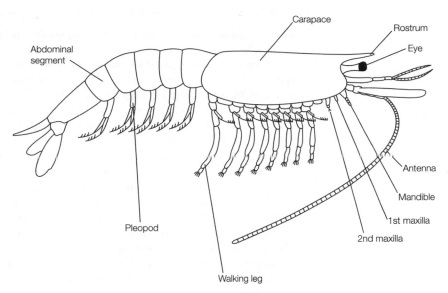

Fig. 3.   *Body plan of a malacostracan crustacean, e.g. a shrimp.*

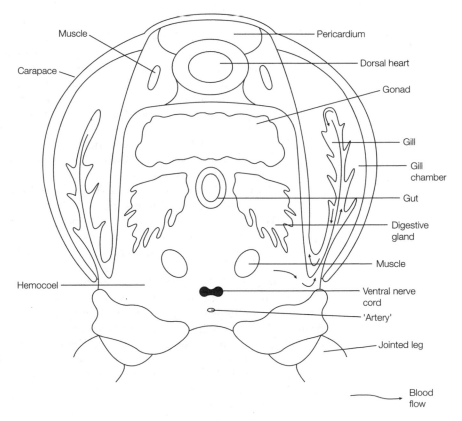

Fig. 4.   *Cross-section through the thorax of a crayfish.*

Many crustaceans are parasites. Six major classes are noted here:

(1) **Class Branchiopoda.** These free-living forms often possess a **bivalved cara-pace** which also encloses a dorsal brood pouch for eggs. The large, biramous antennae are used for swimming. **Parthenogenetic ('virgin-birth') reproduction** is common, with the production of males at only certain times of the year. An example is *Daphnia pulex* (water-'flea').

(2) **Class Ostracoda.** These are rapidly swimming aquatic crustaceans which possess a carapace which encloses the body and the head. An example is *Cythereis* sp. (seed-shrimp).

(3) **Class Copepoda.** The copepods include free-swimming aquatic forms and also parasites which live on fish gills, often appearing as small worms. They are **very common in plankton**. An example is *Cyclops* sp.

(4) **Class Branchiura.** The **fish-louse** is an ectoparasite on fishes: it swims using its thoracic appendages. An example is *Argulus* sp.

(5) **Class Cirripedia.** The **barnacles** have adults which are sessile or parasitic. The shell (or carapace) comprises calcareous plates. Six pairs of thoracic appendages form a 'net' to trap food. Parasitic barnacles may be very degenerate. An example is *Balanus* sp. (barnacle).

(6) **Class Malacostraca.** This class comprises most of the **larger crustaceans** (*Fig. 4*) found in the sea, in freshwater and on land. The body has 20 segments (six in the head, eight in the thorax and six in the abdomen), There are several super-orders, two of which are noted here.

  (i) *Super-order Pericarida.* This has members with a reduced carapace and a brood pouch (resulting in a frequent absence of larval stages). Included among the Pericarida are the **Isopoda** whose terrestrial forms include the **woodlice.** Although described as terrestrial, these crustaceans need to moisten their gill-books with water. The **Amphipoda** include the laterally compressed freshwater shrimps with uniramous thoracic appendages. An example is *Oniscus* sp. (wood-louse).

  (ii) *Super-order Eucarida.* The Eucarida have a carapace which covers the thorax; the eyes have stalks. There is usually a swimming, larval stage. **Euphausids** are important in the food chain as **krill.** The **Decapoda** include shrimps, lobsters and crabs: there are some fresh-water and a few land forms, although they return to the sea to breed. The elongated lobster body format contrasts with that of the crab where the reduced abdomen is tucked under the widened thorax. Examples are *Homarus* spp. (lobster) and *Cancer* spp. (crab).

Sub-phylum Uniramia (millipedes, centipedes, insects)

This enormous sub-pylum contains well over one million species Almost all uniramians are terrestrial (although there are many insects which re-invaded fresh water secondarily). The exoskeleton contains tanned protein. The appendages are uniramous (i.e. have one lobe), the most anterior being the antennae. The mandible bites with the tip. The walking legs lie behind the head. In insects, there is a sharp division between the thorax and the abdomen, but in millipedes and centipedes there are many similar segments. Insects have developed powered flight.

(1) **Class Diplopoda.** The millipedes have between about 25 and 100 segments, each with two pairs of legs. The head has a pair of antennae and two pairs

of mouthparts. Respiration is by tracheae. They are herbivorous burrowers or live in leaf mold. An example is *Spirobolus* sp.

(2) **Class Chilopoda.** Centipedes, usually have fewer segments than the millipedes. The head bears one pair of antennae and three pairs of mouthparts; the next pair of segments has poison claws. Each of the following segments has a pair of walking legs. Respiration is by tracheae. Centipedes are active, fast-moving carnivores living in leaf litter, etc. An example is *Lithobius* sp.

(3) **Class Insecta.**

(a) *The subclass Apterygota.*

This (sometimes erected as a class in its own right) includes the wingless **springtails and silverfishes**. Like all insects, apterygotes have six legs, although there is some debate about whether they originated from a uniramian ancestor separately from the winged insects.

(b) *The subclass Pterygota.*

This contains the 'true' or winged insects. The body is divided into a **head, thorax and abdomen** (*Fig. 5*). The thorax has three segments, each of which bears a pair of legs. Modern insects normally have **four wings**, outgrowths of the body wall; there is a pair of wings on each of the middle and hind segments of the thorax (the mesothorax and metathorax), although the mesothoracic wings may be modified to form 'covers' (elytra) in beetles, and the metathoracic wings may be reduced to 'balancers' (halteres) in flies, or the wings may be secondarily lost (as in some ant generations). (Palæozoic fossil insects are known with a third pair of wings on the anterior prothorax.) The head has a pair of **antennae** and **three pairs of mouthparts**, often greatly adapted to feeding habit and diet. The abdomen lacks legs, but reproductive appendages may be present terminally.

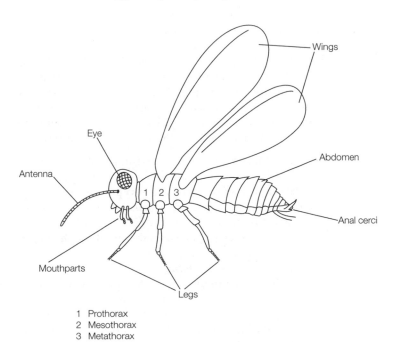

1 Prothorax
2 Mesothorax
3 Metathorax

Fig. 5. Body plan of a typical insect.

Hemimetabolous insects have a series of juvenile stages (as in a locust) in which each instar resembles the adult except for the absence of wings. Holometabolous insects have a larval stage in their life histories which may be very different from the adult (e.g. a maggot or a caterpillar): metamorphosis may be profound, accomplished within a pupa or chrysalis.

Insects breathe through an extensively branching system of tracheal air-ducts lined with chitin (except at the far ends, the tracheoles). In larger insects, air circulation is aided by compressible air-sacs. An open, low-pressure vascular system is present, linked to the hemocoel as the main body cavity: the 'blood' is hemolymph. A dorsal tubular heart works on a suction-pump principle.

Two super-orders are recognized, the exopterygota and the endopterygota.

- **Super-order Exopterygota.** These are hemimetabolous. They include the orders Orthoptera (grasshoppers, crickets and locusts), Dictyoptera (cockroaches and mantises), Odonata (dragonflies and damselflies), Ephemeroptera (mayflies), Dermaptera (earwigs), Isoptera (termites) and Hemiptera (aphids and bugs), together with numerous small orders. An example is *Locusta migratoria* (locust).
- **Super-order Endopterygota.** These are holometabolous. They include the orders Lepidoptera (butterflies and moths), Trichoptera (caddis-flies), Diptera ('true' flies), Siphonaptera (fleas), Coleoptera (beetles, the largest insect order with over 400 000 species) and Hymenoptera (bees, wasps and ants), together with several small orders. Examples include *Drosophila melanogaster* (fruit-fly) and *Apis mellifera* (honeybee).

Sub-phylum Chelicerata (horseshoe crabs, spiders, mites, ticks, scorpions)

There are about 60 000 species of chelicerates. Most are terrestrial, but *Limulus* and a few other related species are marine. Most chelicerates have six pairs of appendages, of which the first two differ from each other: the chelicerae and the pedipalps respectively. There are four pairs of walking legs, one or more with jaw-like gnathobases. The appendages are primitively biramous but are usually uniramous in land species There are no antennae. The body is divided into a prosoma and an opisthosoma. There are three classes.

(1) **Class Merostomata.** The horseshoe crab has a dorsal carapace and a long, pointed telson (tail); horseshoe crabs are marine and are of ancient lineage: Palæozoic Merostomata, the eurypterids, were abundant and attained 3 m in length. An example is *Limulus polyphemus* (horseshoe crab).

(2) **Class Arachnida.** There are about 50 000 species of arachnids. Typically they are terrestrial and breathe air through lung-books or tracheae. The four best-known orders are:

(a) *Scorpionida (Scorpiones).*

In the scorpions (*Fig. 6*), the abdominal opisthosoma is divided into a broader anterior part and a narrow tail bearing a stinging poison gland. The chelicerae are short and powerful, and the pedipalps have large claws. Respiration is by four pairs of lung-books. Scorpions are active predators. An example is *Androctonus australis* (N. African scorpion).

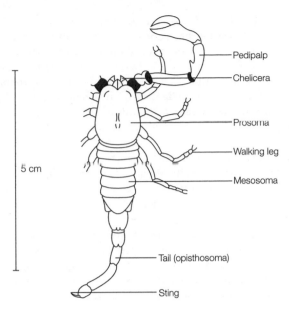

Fig. 6.   A scorpion.

(b) *Araneae*.

The **spiders** (*Fig. 7*) have a spheroidal opsithosoma attached to the anterior prosoma by a narrow waist. The chelicerae have poison glands, whereas the leg-like pedipalps are sensory. Respiration is by lungbooks and/or tracheae. **Spinnerets** (adapted appendages) produce gossamer silk from the opisthosoma: the silk is used for webs to catch prey, for cocooning eggs, etc. All spiders are carnivorous, sucking the digested juices from their prey (which they chase or trap). An example is *Latrodectus mactans* (black widow spider).

(c) *Opiliones (= Phalangida)*.

The **harvestmen** lack the waist of spiders and have longer legs. Most are carnivores, but a few eat carrion or are omnivorous. An example is *Leiobunum* sp.

(d) *Acarina*.

Ticks and mites have an unconstricted junction between the prosoma and the opsithosoma. The pedipalps may be clawed or leg-like; the

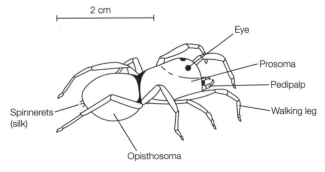

Fig. 7.   A spider.

chelicerae may be clawed. Tracheae and spiracles are often present. Most acarines are very small, living in a variety of habitats from leaf litters to as parasites on animals. Both herbivores and carnivores are known. An example is *Dermacentor variabilis* (tick).

(3) **Class Pycnogonida.** The **pycnogonids or sea-spiders** are problematically related to the Arachnida. Sea-spiders have a reduced abdomen, no special respiratory or excretory apparatus and four, five or six pairs of walking legs. An example is *Nymphon rubrum* (sea-spider).

Related phyla

Phylum Pentastomida (tongue-worms)

The tongue-worms (about 90 species) are endoparasites in the respiratory tracts of carnivorous vertebrates, especially reptiles. The sexes are separate, and mating occurs in the 'final' host. After internal fertilization, the females lay many thousands of shelled eggs which are voided via nasal secretions, saliva or feces. The eggs are eaten by a vertebrate intermediate host and hatch into a larva with four–six jointed, arthropod-like limbs. The intermediate host is eaten by the final host and the larva is freed to migrate to the respiratory tract. Like arthropods, pentastomids have a ventral nerve cord, cerebral ganglia and a hemocoel. An example is *Linguatula serrata*.

Phylum Tardigrada (tardigrades)

The tardigrades or water-bears (about 500 species, <1 mm in length) live in damp intertidal zones, on the edges of freshwater habitats, and on the surface water films of lichens and bryophytes (mosses and liverworts). The sexes are separate. Internal fertilization occurs and the female lays eggs which hatch after 14 days into a juvenile. The juvenile molts to form an adult. Parthenogenesis may be common, especially as some moss-dwelling species seem to have no male individuals. Cryptobiosis – a period of suspended animation when the tardigrades retract their legs, lose water and contract, or very slow metabolism may be seen, which is advantageous when habitats dry out or otherwise become hostile (e.g. extreme cold). Tardigrades have short bodies and four pairs of clawed, short legs. An example is *Echiniscus blumi*.

A12 PHYLUM ECHINODERMATA

Classification: Kingdom Animalia, Sub-Kingdom Metazoa, Phylum Echinodermata

Key Notes

| Features | *Type of organism*: marine, pentaradiate (five-rayed) animals. They show a deuterostome pattern of development. Examples include feather-stars, sea-lilies, starfishes, brittlestars, urchins, sand-dollars and sea-cucumbers. |

Type of organism: marine, pentaradiate (five-rayed) animals. They show a deuterostome pattern of development. Examples include feather-stars, sea-lilies, starfishes, brittlestars, urchins, sand-dollars and sea-cucumbers.

Body plan: early in development they are bilaterally symmetrical, unsegmented, triploblastic coelomates on which is imposed a characteristic pentaradiate symmetry about the oral–aboral axis. The coelom has several sets of cavities, including a 'water vascular system' and main perivisceral cavity.

Feeding: food is obtained by predation, scavenging or suspension feeding. They possess a complete gut with a mouth and anus.

Locomotion: tube-feet associated with the water vascular system are used in some classes; muscular movement of arms is associated with spines in others.

Skeleton: internal skeleton of calcareous ossicles; tube-feet provide a localized hydrostatic skeleton.

Respiration and vascular system: respiration is through the body surface (including infoldings and outfoldings of the body wall). There is no true blood vascular system: the water vascular system, with a fluid composition similar to coelomic fluid, may have respiratory and transport function in some species.

Osmoregulation and excretion: no specialized excretory organs: wastes are voided through the body surface.

Co-ordination: no cephalization; they have a nervous system of rings, radial nerves and nerve nets.

Reproduction: asexual reproduction is common; sexual reproduction with external fertilization (sexes separate). A deuterostome pattern of embryonic development is shown; eggs hatch to produce ciliated, planktonic larvae which are bilaterally symmetrical; pentaradiate symmetry is acquired as the echinoderm grows.

Sub-phyla

Pelmatozoa:
- **Class Crinoidea:** sea-lilies and feather-stars are attached by a stalk joining the aboral apex of the body at some stage of development.

Eleutherozoa (all lack a stalk):
- **Class Asteroidea:** starfishes with arms not set off from the oral disk; use tube-feet for locomotion.
- **Class Ophiuroidea:** brittlestars with arms set off sharply from the oral disk.
- **Class Echinoidea:** urchins lacking arms.
- **Class Holothuroidea:** sea-cucumbers elongated along the oral–aboral axis.

Related phyla	*Chaetognatha*: arrow-worms; bilaterally symmetrical active swimmers with grasping spines on the head.

Chaetognatha: arrow-worms; bilaterally symmetrical active swimmers with grasping spines on the head.
Lophophorate phyla: probably deuterostomes with a three-part body; they possess a lophophore used for feeding and respiration.

- **Phylum Bryozoa:** colonial moss animals with retractable lophophores.
- **Phylum Brachiopoda:** lampshells with two-valved shells: lophophore within the shell.
- **Phylum Phoronida:** sedentary marine worms with lophophores.

Related topics

Body plans and body cavities (B1)	Relationships between phyla (B5)
Symmetry in animals (B2)	Evolutionary origins of the
Protostomes and deuterostomes (B4)	Chordata (B7)

Features

Type of organism
There are about 6000 species of living echinoderms, and many fossil species are known. They are **all marine**, although some live in intertidal zones.

Body plan
Echinoderms are **bilaterally symmetrical, unsegmented, triploblastic coelomates**. They lack segmentation. The body is usually **secondarily pentaradially (five) symmetrical** (imposed on the early larval bilateral symmetry) about an axis from the mouth to the aboral surface (*Figs 1–5*). A tertiary bilateral symmetry may, in turn, be imposed on the pentaradiate symmetry, as in heart urchins.

Feeding
Simple organic molecules can be absorbed directly through the epidermis, but most food is obtained by predation, scavenging or suspension feeding. The gut is usually complete, with a mouth and an anus (although in a few species the anus is secondarily lost).

Locomotion
Locomotion is effected in many forms (e.g. starfishes) by **tube-feet**. The tube-foot is usually bottle-shaped. It has a central, fluid-filled cavity, and shape changes are effected by antagonistic circular and longitudinal muscles. An **ampulla** (bulb) at the proximal end of the tube-foot acts as a compensation chamber for fluid when the elongated (externally visible) part of the foot contracts. The end of the foot may have a sucker, and the foot can be used to haul the echinoderm along: thus the tube-foot has a local hydrostatic skeleton and is used primarily for locomotion in many groups of echinoderms. Muscular movement of arms, aided by spines, effects movement in brittlestars.

Skeleton
The surface of the echinoderm has a delicate epidermis which covers an **endoskeleton of calcareous plates**: these may move or be fixed, and may bear spines. Pedicellariae are minute, jaw-like appendages found on the body walls of starfishes. Perforation in some plates allows tube-feet to project through.

Respiration and vascular system
Respiration is generally through the body surface, but may be effected particularly by papillae, tiny gills, or by the tube-feet. Sea-cucumbers have cloacal

filaments called respiratory trees. There is no true blood, but a **'water vascular system'**, which radiates into the arms and connects with the tube-feet, contains a fluid similar in constitution to coelomic fluid. In some species the coelomic fluid may serve a respiratory or circulatory function. Amoebocytes have a phagocytic, defensive function (arguably homologous to vertebrate macrophages), and cells morphologically and histochemically similar to lymphocytes have been described.

Osmoregulation and excretion
Wastes are generally voided via the body surfaces: specialized excretory organs are not found.

Co-ordination
There is **no cephalization**; the nervous system is rather simple, consisting of a circumoral nerve ring with radial nerves extending from it to a nerve net.

Reproduction
In many echinoderm species reproduction can be effected asexually by fission. For sexual reproduction, the sexes are usually separate with external fertilization taking place. Development follows the **deuterostome** pattern: radial cleavage and the mouth developing from an opening separate from the blastopore. The ciliated echinoderm larva is bilaterally symmetrical, growing and (in many species) passing through several stages before metamorphosis to an adult. Many echinoderms exhibit impressive regeneration capacities. Similarities in embryonic and larval development point to links between the echinoderms and the chordates.

Sub-phyla

There are two echinoderm sub-phyla, the second containing four classes.

Sub-phylum Pelmatozoa
The sub-phylum Pelmatozoa contains the single class **Crinoidea**, the sea-lilies and feather-stars. These echinoderms are attached to the substratum or the colony, either permanently or during development, by a **stalk** from the aboral apex of the body. The oral surface faces upwards. The arms are sharply distinct from the oral disk and bear pinnules. Tube-feet are used for filter-feeding and for respiration. The sea-lilies are sessile, stalked forms (*Fig. 1*); feather-stars swim. Many fossil forms are known, the earliest from the Cambrian Burgess shale (about 530 million years B.P.). An example is *Antedon* sp. (feather-star).

Sub-phylum Eleutherozoa
The Eleutherozoa **lack a stalk**; there are four classes:

(1) **Class Asteroidea.** The **starfishes** are typically pentaradiate with the **arms not set off distinctly from the oral disk** (*Fig. 2*). They use their **tube-feet as extensible, flexible, sucking devices** for locomotion. Many feed on bivalve molluscs or suspension-feed using cilia. Pedicellariae help to kill small organisms which settle on the starfish surface. An example is *Asterias rubens* (starfish).

(2) **Class Ophiuroidea.** Brittlestars are also typically pentaradiate but their **arms are sharply set off from the oral disks** (*Fig. 3*); the arms contain articulating ossicles or 'vertebrae', spines give a spiked appearance. Brittlestars feed on small particles. An example is *Ophiura* sp. (brittlestar).

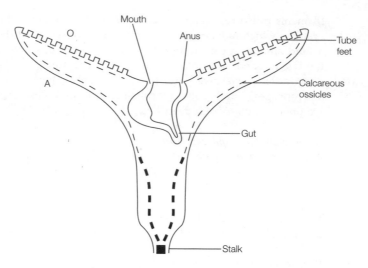

Fig. 1.   A pelmatozoan (sea-lily). O, oral surface; A, aboral surface.

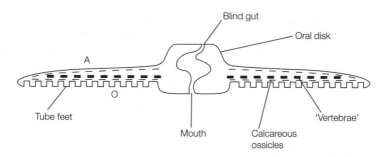

Fig. 2.   An asteroidean (starfish). O, oral surface; A, aboral surface.

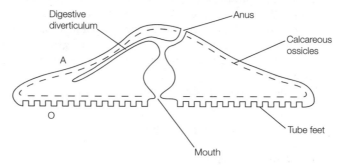

Fig. 3.   An ophiuroidean (brittlestar). O, oral surface; A, aboral surface.

(3) **Class Echinoidea.** The **sea-urchins, sand-dollars and their allies** are pentamerous but **lack arms**: the oral surface extends over most of the **globose body** whose skeleton forms a **test** (*Fig. 4*). Locomotion is by spines (which can also be long, protective and poisonous) and/or tube-feet. Five teeth and an **Aristotle's lantern** allow the urchin to scrape algae off rocks. In heart-urchins and sea-potatoes, **bilateral symmetry is**

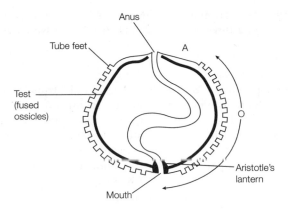

Fig. 4.   An echinodean (sea-urchin). O, oral surface; A, aboral surface.

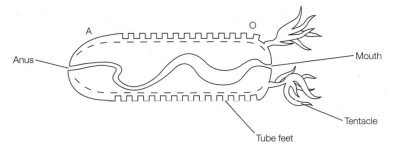

Fig. 5.   A holothuroidean (sea-cucumber). O, oral surface; A, aboral surface.

superimposed on the pentaradiate symmetry: the mouth is 'anterior' and the anus is 'posterior'. Heart-urchins burrow and lack the Aristotle's lantern. An example is *Echinus esculentus* (edible sea-urchin).

(4)   **Class Holothuroidea.** The **sea-cucumbers** and *bêches de mer* are elongated along the oral–aboral axis so that they **lie on their 'sides'** (*Fig. 5*). The **oral surface is greatly expanded**. The endoskeleton is reduced within the muscular body walls. There are many tube-feet, extended to form 10 branched tentacles around the mouth. They suspension-feed or are selective deposit feeders. An example is *Cucumaria frondosa* (North Atlantic sea-cucumber).

Related phyla   *Phylum Chaetognatha (arrow-worms)*

There are about 100 species of **arrow-worms**, a phylum well preserved in the fossil record from about 500 million years B.P. These deuterostome coleomates possess bilateral symmetry and have a streamlined appearance (*Fig. 6*). Most are active marine swimmers although a few exist on the sea bottom. Their small size allows gas exchange and excretion to occur through the body wall. The nervous system is simple, although cephalization is seen: there are two compound eyes. Characteristic of chaetognaths are **grasping spines on the head**, used to seize prey. Chaetognaths are important components in plankton. They are hermaphrodites and reproduce sexually: fertilization is internal, miniature adults hatching directly from eggs shed into the water. An example is *Sagitta elegans*.

The Lophorate phyla
The lophophorates (bilaterally symmetrical, triploblastic coelomates) are **probably deuterostomes**, although the mouth often forms from or near the blastopore. The body comprises three parts: a **prosome, mesosome and metasome**, normally each with a separate coelomic compartment. There is a complex, U-shaped gut and a tough outer cuticle.

Diagnostic of lophophorates is a **lophophore**. This is a circular or U-shaped ridge bearing one or two rows of hollow, ciliated tentacles. The lophophore is used for feeding and for respiration. Adult lophophores are sessile: the lophophore is used to capture plankton.

There are three, related phyla:

(1) **Phylum Bryozoa (= Ectoprocta; moss animals).** Bryozoans are colonial animals living in a test secreted by the body wall; most are marine. The indiviuals within the colony (which is formed by asexual reproduction) are connected by tissue strands along which substances can move. The lophophore is retractable. Individuals in a colony, which may comprise millions of animals, may exhibit differentiation according to role, for example feeding, reproduction (cf. cnidarian polyp colonies). Sexual reproduction is also seen, with internal fertilization of eggs. An example is *Plumatella repens*.

(2) **Phylum Brachiopoda (lampshells).** This ancient phylum has animals with **two-valved shells** which can be closed to protect the soft brachiopod body (*Fig. 7*). The shell **valves are dorsal and ventral** (rather than lateral as in bivalve molluscs). The lophophore lies within the shell: cilia beat to draw water into the shell cavity, and food is caught on the lophophore. Brachiopods usually lie buried in sand or mud, the anterior protruding above the substratum and being supported by a worm-like stalk (*Fig. 7*). The phylum is well-known to palaeontologists: over 25 000 fossil species are known (compared with <400 species alive today), mostly from the Palæozoic and the Mesozoic. An example is *Lingula* sp.

(3) **Phylum Phoronida.** This small phylum (15 species) is made up of sedentary marine worms living in sand or mud, or attached to rocks. They live in secreted, chitinous tubes. Cilia propel water down through the lophophore where food is trapped prior to transport to the mouth. An example is *Phoronis australis*.

A13 PHYLUM HEMICHORDATA

Classification: Kingdom Animalia, Sub-Kingdom Metazoa, Phylum Hemichordata

Key Notes

Features	*Type of organism*: marine worms or small sessile organisms, probably closely related to the chordates. They show a deuterostome pattern of development. Examples are acorn-worms. *Body plan*: triploblastic, bilaterally symmetrical coelomates. They have an unsegmented body which is divided into three regions. *Feeding*: pterobranchs filter-feed with a lophophore-like organ; enteropneusts are detritus feeders, usually with gill-slits present in the pharynx. *Locomotion*: sessile, or muscular burrowers. *Skeleton*: stiffening rod beneath the anterior diverticulum of the pharynx, formally known as a stomatochord. *Respiration and vascular system*: gas exchange occurs over the gill-slits and across body surfaces. There is an open circulatory system with contractile dorsal and ventral vessels. *Osmoregulation and excretion*: most wastes are lost by diffusion through body surfaces. *Co-ordination*: mid-dorsal (in parts hollow) and mid-ventral nerve cords; nerve net beneath the epidermis. *Reproduction*: enteropneusts with separate sexes and many pairs of lateral gonads. They show external fertilization and a deuterostome development pattern. They produce tornaria larvae – pterobranchs reproduce sexually and by asexual budding.
Classes	*Pterobranchiata*: sessile, marine forms, usually colonial with lophophore-like, ciliated tentacles mounted on a shield. There are either two or no pairs of gill-slits. *Enteropneusta*: burrowing acorn-worms; proboscis, collar and long, worm-like body with numerous gill-slits.
Related topics	Phylum Chordata (A14) Protostomes and deuterostomes (B4) Body plans and body cavities (B1) Relationships between phyla (B5)

Features

Type of organism

There are about 90 species of hemichordates. All are marine and live in the open ocean or in muddy sediments.

Hemichordata were once included in the phylum Chordata, and were grouped with the invertebrate chordates as **'protochordates'**: like chordates, they have a dorsal nerve cord, often hollow, developed from the dorsal epidermis. The longitudinal, stiffening **stomatochord** is not now thought to be the **homolog of the notochord**, and this distinction justified the separation of the Hemichordata from the Chordata.

Fossil hemichordates are known from Ordovician times (450 million years B.P.), and forms similar to hemichordates have been described from the Cambrian Burgess shale. The extinct **graptolites** *may* be near relatives.

Body plan

Hemichordata are triploblastic, bilaterally symmetrical coelomates with unsegmented bodies divided into three regions. They usually possess **pharyngeal gill-slits**. The body is divided into a prosome, mesosome and metasome, each with a coelomic cavity (cf. lophophorate phyla).

Feeding

Pterobranchs filter-feed using a lophophore-like organ; enteropneusts are detritus feeders, using the gill-slits to void residual waste.

Locomotion

Pterobranchs are sessile; enteropneusts are burrowers.

Skeleton

Enteropneusts possess a rod ventral to the anterior diverticulum of the pharynx, which has a stiffening function. Formerly known as a stomatochord, it was once thought to be the homolog of the chordate notochord.

Respiration and vascular system

Respiratory gas exchange is across the gill-slits which have vascularized regions beneath their surfaces, and across the general body surface; there is a low-pressure circulatory system with contractile ventral and dorsal vessels linked by open sinuses. In the prosome is a pulsatile 'heart' which pumps blood through the open circulatory system.

Osmoregulation and excretion

Nitrogenous wastes are voided by diffusion through the body surfaces.

Co-ordination

There are normally **mid-dorsal and mid-ventral nerve cords,** linked by nerve rings and sub-epidermal nerve nets.

Reproduction

The hemichordates have separate sexes; fertilized eggs subsequently follow a **deuterostome** pattern of development. In enteropneusts, development may be via a tornaria larva, or directly into a miniature adult. Sexual reproduction by fragmentation also occurs. Pterobranchs reproduce sexually and asexually by budding.

Classes

Class Pterobranchiata

The pterobranchs are small (<8 mm) marine forms (*Fig. 1*), usually **colonial**, living in **secreted tubes**. They have **lophophore-like, ciliated tentacles mounted on a shield**, and feed on small, suspended particles. The gut is U-shaped: **two, or no pairs of pharyngeal gill-slits** may be present. Examples include *Rhabdopleura* sp.; *Cephalodiscus* sp.

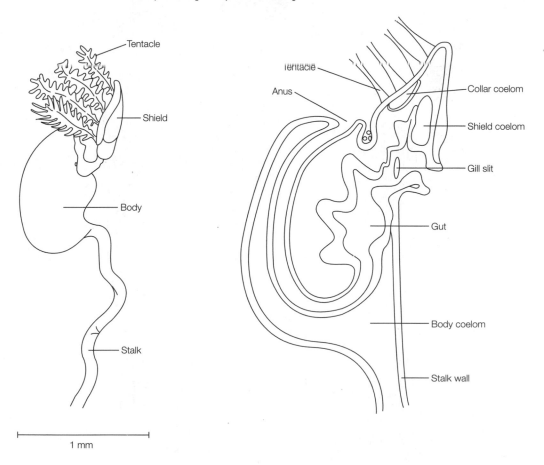

1 mm

Fig. 1.   A pterobranch, e.g. Cephalodiscus *sp.*

Class Enteropneusta

The marine **acorn-worms** are **solitary** (*Fig. 2*). They have a **proboscis** instead of a shield and tentacles, a **collar**, and a long, vermiform (worm-like) body containing a straight gut with **numerous gill-slits**. Examples are *Dolichoglossus* sp.; *Balanoglossus* sp.

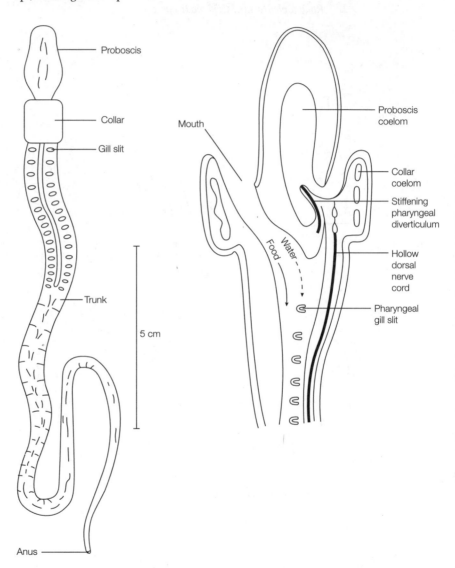

Fig. 2.   An enteropneustan (acorn-worm), e.g. Dolichoglossus *sp.*

A14 PHYLUM CHORDATA

Classification: Kingdom Animalia, Sub-Kingdom Metazoa, Phylum Chordata

Key Notes

Features

Type of organism: large and diverse phylum, mainly comprising vertebrates; marine, freshwater and terrestrial species. Several groups are capable of aerial flight. They show a deuterostome pattern of embryonic development. Examples include sea-squirts, salps, amphioxus, fishes, amphibians, reptiles, birds and mammals.

Body plan: bilaterally symmetrical, triploblastic, segmented coelomates. They possess, at some stage in their development, a dorsal, hollow nerve cord, a notochord, pharygeal gill-slits and (usually) a post-anal tail. There are segmented muscles in the body.

Feeding: invertebrate chordates filter-feed. Most vertebrates are macrophagous feeders using toothed jaws. The gut is complete, with a mouth and anus. The invertebrate chordate pharyngeal endostyle is the homolog of the vertebrate thyroid gland.

Locomotion: mostly active swimmers; segmental muscles act against an antitelescopic notochord or vertebral column; stabilizing and controlling paired fins in fishes evolve into paired limbs of land vertebrates (sometimes further adapted for flight).

Skeleton: they possess an endoskeleton of cartilage or bone. The characteristic notochord is replaced by intersegmental vertebrae in vertebrates.

Respiration and vascular system: aquatic breathers use pharyngeal gill-slits with gills; terrestrial forms respire using lungs. Usually a closed, high-pressure vascular system with a ventral heart.

Osmoregulation and excretion: invertebrate chordates are without discrete excretory organs or have solenocytes; vertebrates possess segmental or nonsegmental kidneys.

Co-ordination: hollow dorsal nerve cord diagnostic of the phylum; the anterior part expands in vertebrates to form the brain.

Reproduction: sexual reproduction normal, although asexual budding is seen in some urochordates.

Sub-phyla: Urochordata, Cephalochordata and Craniata (vertebrata).

Sub-phylum Urochordata

The larva has characteristic chordate features, which are usually lost at metamorphosis when the pharynx enlarges to form an elaborate filter-feeding branchial basket. The body is enclosed in a test or tunic. Urochordates are hermaphrodite and asexual budding is seen.

Class Ascidiacea: sessile sea-squirts.
Class Larvacea: retain some larval features into adulthood.
Class Thaliacea: salps, motile after metamorphosis.

| Sub-phylum Cephalochordata | Amphioxus. Motile, with asymmetrical body architecture. The notochord extends to the extreme anterior end of the body. |

| Sub-phylum Craniata (=Vertebrata) | The dorsal nerve cord expands anteriorly to form a brain enclosed within a cranium. The notochord is usually replaced by cartilaginous or bony intersegmental units (vertebrae). |

Superclass Agnatha: jawless fishes (e.g. modern hagfishes and lampreys).
Superclass Gnathostomata: jawed vertebrates in which the first gill-bar has become 'wrapped round' the mouth to form jaws.

- **Class Acanthodii:** extinct fishes with paired fins.
- **Class Placodermi:** extinct shark-like fishes with bony armor plates.
- **Class Chondrichthyes:** fishes with cartilaginous skeletons; include **Elasmobranchii** (sharks and rays) and **Holocephali** (rat-fishes).
- **Class Osteichthyes:** bony fishes (>30 000 species). Include **Actinopterygii** (ray-finned fishes, including **Teleostei**, the largest group) and **Sarcoptyergii** [fleshy-finned fishes, including lungfishes (**Dipnoi**) and extinct groups from which land vertebrates evolved].
- **Class Amphibia:** forms with aquatic larvae which undergo metamorphosis to more terrestrially adapted forms. The skull does not articulate with the pectoral girdle.
- **Class Reptila:** diverse class with amniote egg; the articular bone of the lower jaw articulates (hinges) with the quadrate bone of the skull.
- **Class Aves:** birds: a homogenous group with amniote eggs, homoiothermy and feathers.
- **Class Mammalia:** possess amniote egg (usually viviparous); fur present; suckle young with milk from mammary glands.

Related topics

Body plans and body cavities (B1)
Skeletons (B3)
Protostomes and deuterostomes (B4)
Relationships between phyla (B5)
Neoteny and pedogenesis (B6)

Evolutionary origins of the Chordata (B7)
Sections on physiological systems [e.g. Nitrogenous excretion (C17)]
Development and birth (D2)
Metamorphosis (D4)

Features

Type of organism
The Chordata is a very large and diverse phylum which has been studied extensively, mainly because it includes the **vertebrates**. **Invertebrate chordates** exist, however, and these are **important in evolutionary terms**. These are sometimes called *protochordates*, a term which formerly included the Hemichordata too. There are about 45 000 species of chordates, occupying marine, fresh and brackish water and terrestrial habitats. Several groups have evolved flight.

Body plan
Chordates are **bilaterally symmetrical, triploblastic, segmented coelomates** which demonstrate a **deuterostomic** pattern of early embryonic development.
   At some stage in their development they possess: a **dorsal tubular nerve cord**, formed from an infolding of a strip of dorsal ectoderm; a **notochord**; **pharyngeal gill-slits**; and, usually, a **post-anal tail**. Segmented muscles are found in the body.

Feeding
Invertebrate chordates are **filter-feeders**: food particles are trapped in mucus on elaborately expanded pharyngeal gill-slit systems and wafted into the gut by cilia; modern **jawless vertebrates** are **semi-parasites** on jawed fishes; most **jawed vertebrates** are herbivorous or carnivorous, **macrophagous feeders**. Teeth inserted into vertebrate jaws are almost universal. The gut is complete with a mouth and an anus. An **endostyle** along the floor of the pharynx in invertebrate chordates is the homolog of the vertebrate **thyroid** gland.

Locomotion
Most chordates are **active swimmers**, using muscles which act against the **anti-telescopic notochord or vertebral column**. (Many urochordates metamorphose to form sessile adults.) The evolution of median and paired fins in fishes facilitates control of pitching, yawing and rolling. In land vertebrates, the paired fins have evolved to form **jointed limbs** which act as levers against the substratum: these may be secondarily lost (as in snakes). A number of chordate groups have evolved flight and others have resumed an aquatic habitat, usually using modified paired limbs.

Skeleton
Chordates are characterized by an **endoskeleton**; in the vertebrates the endoskeleton is made of **cartilage**, normally replaced by **bone** (consisting primarily of hydrated calcium phosphate and protein). The **notochord** forms as an antitelescopic, cartilaginous rod dorsal to the gut and ventral to the nerve cord. It contains a gelatinous matrix surrounded by tough connective tissue. In vertebrates, it is partly or completely replaced by **vertebrae** which develop **intersegmentally** and surround the nerve cord.

Respiration and vascular system
The pharyngeal gill-slits, used for filter-feeding in invertebrate chordates, are used for **respiration** in the more advanced aquatic forms but are present only transiently in the embryo in land vertebrates. **Gills** develop in fishes and are present in amphibian larvae. Land vertebrates respire using **lungs** which develop as pouches from the gut (and are the homolog of the fish swim-bladder). Most chordates have a **high-pressure, closed circulation with a ventral heart**. A dorsal vessel conveys blood backwards while a ventral vessel conveys it forwards.

Osmoregulation and excretion
The characteristic excretory organs of vertebrates are the segmental **kidneys**, replaced in reptiles, birds and mammals by nonsegmental kidneys; invertebrate chordates rely on simple diffusion of waste substances (urochordates) or groups of **solenocytes** (cephalochordates).

Co-ordination
The **dorsal hollow nerve cord**, present at some stage during the chordate's life history, is characteristic of this phylum. In metamorphosed urochordates the nerve cord is lost. In vertebrates the anterior end of the nerve cord expands to form a **brain**.

Reproduction

Reproduction is normally sexual, although parthenogenesis is seen in a few species, and some urochordates reproduce asexually by budding to form colonies.

Sub-phyla

There are three chordate sub-phyla:

(1)   Urochordata
(2)   Cephalochordata
(3)   Craniata (= Vertebrata)

The first two are sometimes called the acraniate or invertebrate chordates. The craniates are more popularly termed the **vertebrates**.

Sub-phylum Urochordata (Tunicata)

The 1300 species of urochordates or tunicates have **larvae** (*Fig. 1a*) with the typical **chordate structure** (including a notochord, a dorsal nerve cord and a segmentally muscled tail). However, after metamorphosis, this structure apart from the gill-slits is lost. The tail with its musculature and the notochord disappear, while the nerve cord is reduced to a round ganglion from which peripheral nerves radiate. The pharynx enlarges to form a **branchial basket** with **numerous**

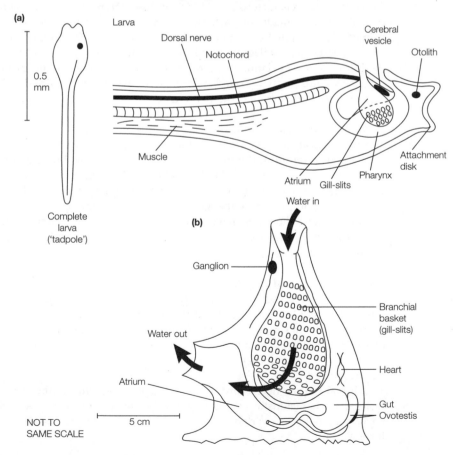

Fig. 1.   (a) A larval tunicate (urochordate) and (b) an adult sea-squirt tunicate, e.g. Ciona *sp.*

ciliated gill-bars covered in mucus. Water is drawn through the mouth, passing through the gill-slits to an atrium. Food particles are filtered out in the mucus wafted by cilia along the gill-bars. They pass to a **dorsal epipharyngeal groove** and thence back to the gut. Ventrally in the pharynx is an **endostyle**, which concentrates iodine and is the homolog of the vertebrate thyroid gland. The body is enclosed in a cellulose **'test'** or **tunic**.

Urochordates have a heart (which can reverse its direction of beat) and blood vessels. Functional lymphocytes are seen, and **rudiments of the vertebrate adaptive immune system** have been described. Blood cells include vanadocytes (containing hemovanadium, although this vanadium-containing pigment has not been demonstrated to have a respiratory function).

Urochordates are usually hermaphrodite with an **ovotestis**. Eggs develop into a 'tadpole' larva which metamorphoses into an adult. Many **multiply asexually by budding** to form a colony with a common exhalant opening from the atrium.

There are three classes: the Ascidacea, the Larvacea and the Thaliacea.

Class Ascidiacea: sea-squirts
Usually **sessile** and bottom-living (*Fig. 1b*). They may be solitary or colonial. Examples include *Ciona intestinalis* and *Botryllus* sp.

Class Larvacea
Forms with a **neotenic retention of some larval features**, such as a tail, into adulthood. *Oikopleura* has an elaborate, secreted 'house' used for filter-feeding. An example is *Oikopleura dioica*.

Class Thaliacea
Salps, solitary or colonial. The test has muscle bands and the atrial opening points posteriorly allowing the salp to **swim by jet-propulsion**. An example is *Salpa* sp.

Sub-phylum Cephalochordata

The Cephalochordata comprise two genera. In *Branchiostoma lanceolatum* (= *Amphioxus lanceolatum*), the lancelet or **amphioxus** (*Fig. 2*), the **notochord persists throughout life**: it **extends to the extreme anterior end**, beyond the hollow nerve cord which lies dorsal to it; the nerve cord does not expand to form a brain and there is **little cephalization**. The animal lies half-buried in the sand, filter-feeding particles from the sea water.

Cephalochordates are small (<5 cm). A specialized feature of the sub-phylum is the **asymmetrical body architecture** whereby segmented muscles and nerves alternate on each side of the body. (The second cephalochordate genus is called *Asymmetron*.) The pharynx is long with many ciliated gill-bars: the cilia generate a current so that water is drawn into the mouth, passes through the gill-slits where food is filtered out, and then enters the atrium and thence out through the atriopore (*Fig. 3*). Mucus is produced by a ventral endostyle (the homolog of the vertebrate thyroid gland) in the pharynx. Similarities with the feeding patterns of urochordates are seen.

Gonads are arranged segmentally: the sexes are separate. Excretion is facilitated by groups of flame cells (solenocytes) similar to those in platyhelminths and polychaete annelids. The blood vessel flow pattern is similar to that in

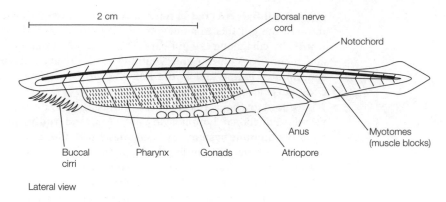

Lateral view

Fig. 2.   A cephalochordate, amphioxus, showing the notochord.

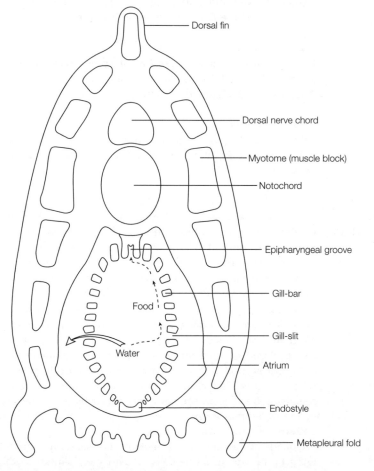

Fig. 3.   Cross-section through the pharynx of amphioxus.

vertebrates, but there are no blood cells nor a heart: the vessel walls are contractile. An example is *Branchiostoma lanceolatum* (amphioxus or lancelet).

Sub-phylum Craniata (= Vertebrata)

Craniates possess the diagnostic chordate features (at some stage in their life histories) of a notochord, hollow dorsal nerve cord, pharyngeal gill-slits and a post-anal tail. All craniates have an anterior expansion of the dorsal nerve cord, the **brain**, contained in a **cranium** (the brain-box or skull) (*Fig. 4*). This **cephalization** is further marked by extensive development of sense organs in the head.

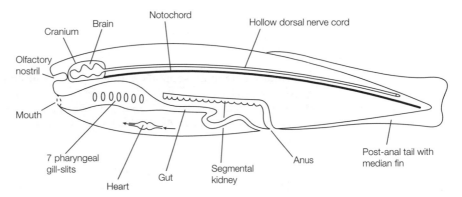

Fig. 4.   A hypothetical primitive craniate (vertebrate).

The **notochord is usually replaced** in part or totally by intersegmental cartilage or bony units, the **vertebrae**. Cartilage is found in some invertebrates such as Mollusca, but **bone** is unique to vertebrates, mainly consisting of hydrated calcium phosphate minerals in a protein matrix.

There is a **ventral heart** behind the head. Blood flows forward ventrally from the heart, then through the gills, and then upwards and backwards dorsally. The excretory organs, the **kidneys**, lie dorsally in the coelom: they comprise thousands of nephron units.

The sub-phylum is normally divided into two superclasses: the agnathans and the gnathostomes.

Superclass Agnatha (jawless vertebrates)
The earliest known vertebrates, the **ostracoderms**, **lacked jaws**: these armored fishes are totally extinct. Modern Agnatha belong to the class **Cyclostomata** (round mouths): this comprises the **lampreys and the hagfishes** (*Fig. 5*). The absence of jaws is diagnostic. Cyclostomes have round, suctorial mouths. They are semi-parasites on larger, jawed fishes to which they attach, rasping at the host flesh with a toothed tongue and horny teeth within the mouth cavity. The sexes are separate. The fertilized eggs of lampreys hatch to form filter-feeding ammocoetes larvae which lie in the mud; after some years they metamorphose into lampreys.

Specialized features of lampreys include the mouth and tongue, the streamlined shape, and pouched gills allowing a tidal flow of water over them.

Primitive features include a persistent notochord, a lack of paired fins, no ducts from the gonads and seven pairs of gills. The skeleton is cartilaginous

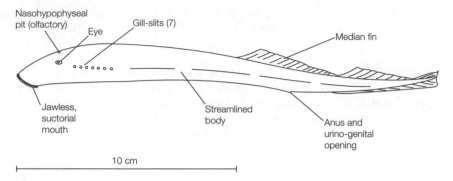

Fig. 5. *A typical agnathan (lamprey).*

and includes a cranium, supports anterior to each gill pouch (gill arches) and peg-like vertebrae.

Degenerate features include reduced eyes and a simple, straight gut. Examples include *Lampetra fluviatilis* (common lamprey) and *Myxine glutinosa* (Atlantic hag).

Superclass Gnathostomata (jawed vertebrates)

The Gnathostomata includes all other vertebrates. In gnathostomes the **first gill-bar** becomes 'wrapped round' the mouth to form the **upper and lower jaws** (*Fig. 6*). This allows feeding on large particles (**macrophagy**), especially when

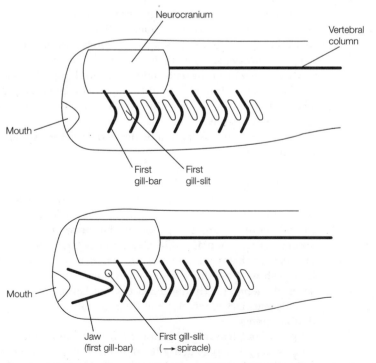

Fig. 6. *A typical gnathosome, showing evolution of jaws in craniates: the first gill-bar becomes the jaws.*

teeth are inserted into the jaws. Eight classes of gnathostome are usually recognized.

(1) **Class Acanthodii.** The classification of the extinct acanthodians, known from the Devonian, is problematical: they exhibit affinities to cartilaginous fishes, to bony fishes and to forms unknown today. They are the **first fishes to have true paired fins** (unlike the paddles of some ostracoderm agnathans) and to possess jaws. The paired fins were attached to the trunk, and there were **intermediate, paired spines**, between the pectoral and pelvis fins, apparently not associated with fin structures. An example is *Climatius* sp.

(2) **Class Placodermi.** The placoderms were Palæozoic shark-like fishes with jaws and an armor of bony plates. There were several subclasses, now all extinct. An example is *Dunkleosteus* sp.

(3) **Class Chondrichthyes.** First known from the upper Devonian, the Chondrichthyes (cartilaginous fishes) have two subclasses. With the exception of only four or five species they are marine. Chondrichthyes have a **cartilaginous skeleton** (which may be secondarily calcified) and scales, identical in structure to teeth (including dentine). Pectoral and pelvic **paired fins** are present. Fertilization is internal, with the females laying large, yolky eggs or exhibiting (ovo)viviparity.

   (a) *Subclass Elasmobranchii.*

   The elasmobranchs include the sharks, dogfishes, skates and rays. The upper jaws move with respect to the cranium; there are 5–7 gill-slit pairs and an anterior spiracle, each opening separately to the exterior. The order **Selachii** includes the sharks, many of which are large predators; the order **Batoidea** includes skates and rays with dorso-ventral flattening and enlarged pectoral fins. Many batoideans are molluscivores with flattened, crushing teeth. Examples include *Carcharodon carcharias* (great white shark); *Manta* sp. (manta ray).

   (b) *Subclass Holocephali.*

   The deep-sea rat-fishes, rabbit-fishes and chimeras may not be closely related to the sharks and rays. Their skeletons, gill and vascular systems, and their nervous systems are different from those of the Elasmobranchii. An example is *Hydrolagua colliei* (rat-fish).

(4) **Class Osteichthyes.** With more than 30 000 species, the Osteichthyes (bony fishes) include more species than all the other vertebrate classes put together. The embryonic cartilage in the skeleton is replaced by bone. Paired fins are attached to limb girdles. The gill-slits open to an opercular cavity which is enclosed within an **operculum**. Pharyngeal pouches from the foregut were probably primitive lungs: these may be modified to form a **swim-bladder**, used for buoyancy. Fertilization of the eggs is external: large numbers of small eggs develop into larvae (fry) which metamorphose into adults.

   There are two subclasses:

   (a) *Subclass Actinopterygii.*

   The Actinopterygii or **ray-finned fishes** have bones and muscles restricted to the bases of the fins which are supported distally by fin rays. The bony scales have a thick layer of enamel. Almost all bony fishes alive today are actinopterygians: three grades or infraclasses are recognized, with progressively more advanced features:

- **Chondrostei.** All early actinopterygian fossils belong to a group of Chondrostei known as the **Palaeonisciformes** (palaeoniscids) which radiated extensively between Carboniferous and Triassic times; the modern Chondrostei are relics and include the bichir, the paddle-fishes and sturgeons. Air-sacs or lungs are present. Examples are *Polypterus ornatipinnis* (bichir) and *Huso huso* (Volga sturgeon).
- **Holostei.** At this grade, morphological body changes allow for more efficient swimming and feeding. Most Holostei died out in the Cretaceous and the Eocene, but survivors include the bowfin and the gar-pike. Examples are *Amia calva* (bowfin); *Lepisosteus osseus* (gar).
- **Teleostei.** Most living fishes are teleosts (about 30 000 species): the grade or infraclass radiated extensively during the early Tertiary. Teleosts have thin, bony scales without enamel, a moveable upper jaw which hinges anteriorly to the skull, a complete vertebral columns which replaces the notochord and a single, dorsal swim-bladder which may or may not be connected to the gut. Teleost classification is complex: many zoologists feel that they show polyphyletic (multiple evolutionary) origins from the Holostei. Examples are *Cyprinus carpio* (carp); *Esox lucius* (pike).

(b) *Subclasss Sarcopterygii.*

The sarcopterygians or **fleshy-finned fishes** have paired fins with an axis of bone and muscle. There are two infraclassses:

- **Crossopterygii.** The main importance of the crossopterygians (the tassel-finned fishes) lies in the fact that from them probably arose the land vertebrates (the **Tetrapoda**). Forms such as *Osteolepis* sp. have similar limb bone and skull bone arrangements to the earliest amphibians. There is only one living species, the **coelacanth**. This ovoviviparous fish (from deep waters off the Comoro islands in the western Indian Ocean) has a primitive blood system and a fat-filled swim-bladder. The living example is *Latimeria chalumnae* (coelacanth).
- **Dipnoi.** The **true lungfishes** survive today in three living species, from Australia, South America and East Africa respectively, being relics of a group abundant in Devonian fresh waters. Dipnoans have lungs and breathe air but they also retain their gills. The African and South American forms are obligate air-breathers, but the Australian lungfish needs oxygenated water to survive. The blood system is, for fishes, sophisticated, with a partially divided heart and with pulmonary veins returning oxygenated blood to the left atrium. Estivation in cocoons the mud is common during drought periods. Male African and South American lungfishes brood eggs which hatch into larvae with external gills. Examples include *Neoceratodus forsteri* (Australian lungfish); and *Lepidosiren paradoxa* (South American lungfish).

(5) **Class Amphibia.** Amphibians arose from the crossopterygian fishes during the early Devonian: these early amphibians were known as **Labyrinthodontia** on account of the architecture of their teeth. Modern forms, collectively known as **Lissamphibia**, have problematical relationships with the labyrinthodonts.

Amphibians typically possess two pairs of legs (not fins), although one or both pairs may be secondarily absent. The pectoral girdle does not connect to the back of the skull; there is a pelvic girdle connected to the vertebral column by sacral vertebrae. The middle ear has a single bone connecting the eardrum to the inner ear.

Amphibian larvae (tadpoles) have external gills which may later be internalized. They develop in water, which may include specialized habitats such as pools in the bases of flowers or a brood pouch in the mother's mouth. **Metamorphosis** is under the control of the **hypothalamus–pituitary–thyroid axis** and a **prolactin–thyroxin balance**. Metamorphosis may be profound: limbs develop and lungs replace gills (although many species also employ cutaneous respiration). Some urodele species become sexually mature without undergoing metamorphosis, exhibiting **neoteny** (retention of larval characters in the adult) and **pedogenesis** (sexual maturity in the larva). Amphibian hearts have two atria and one ventricle.

There are three living orders of Lissamphibia:

(a) *Order Anura.*

The anurans are the **toads and frogs** which in the adult form lack tails. Sexual maturity is always preceded by metamorphosis from a tadpole larva. Examples are *Rana temporaria* (common frog); *Rana catesbeiana* (bullfrog); *Xenopus laevis* (African clawed toad).

(b) *Order Urodela.*

The **newts and salamanders** and their allies have long tails used for swimming. Forms range from those that are totally terrestrial, through **facultative neotenes** which are sexually mature as larvae and only metamorphose in response to drought or thyroxine treatment, through to **perennibranchiates** which are never observed to metamorphose. These neotenic forms retain their gills. Examples are *Triturus cristatus* (crested newt); *Ambystoma mexicanum* (Mexican axolotl).

(c) *Order Apoda.*

The apodans or gymnophionids are **limbless, burrowing amphibians** from the tropics or southern temperate regions; the larval stage may pass within the egg. An example is *Dermophis mexicanus.*

(6) **Class Reptilia.** Reptiles evolved from labyrinthodont amphibians during the late Devonian. The ancestral group were the **Cotylosauria** or 'stem-reptiles'. From this group evolved various taxa including the Chelonia (tortoises and turtles), the Lepidosauria (the tuatara, lizards and snakes), the Synapsida (whose later members included the mammal-like reptiles which gave rise to the mammals), the Archosauria (whose living members include the crocodiles but which once claimed the flying pterosaurs and the Ornithischia and the Saurischia, the two 'dinosaur' orders) and several extinct groups of marine reptiles (such as ichthyosaurs and plesiosaurs).

Like birds and mammals, reptiles have an amniote egg (see Topic D2): the embryo develops an amnion, chorion and allantois in addition to a yolk-sac. The three classes are sometimes grouped as a superclass Amniota. There are no larval stages, and some reptiles (e.g. sea-serpents and adders) are viviparous. The chorioallantois acts as a respiratory organ in the developing, waterproof egg. Gills are absent. Oviparous marine and freshwater species (e.g. turtles and crocodiles) must lay eggs on land.

In reptiles the articular bone of the lower jaw articulates with the quadrate bone of the skull (*Fig. 7*). Reptiles have a more-or-less waterproof

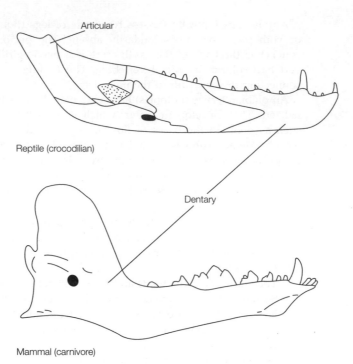

Articular

Reptile (crocodilian)

Dentary

Mammal (carnivore)

Fig. 7.   Comparison of the lower jaws of a reptile (crocodilian) and a mammal (carnivore). In reptiles, the articular component articulates (hinges) with the quadrate bone of the skull; in mammals, the dentary articulates with the squamosal bone of the skull.

skin, usually with scales. The kidney develops nonsegmentally and is known as metanephric (as opposed to the segmental, meso- or opistho-nephric kidneys of fishes and amphibians).

Reptiles reached their peak during the Mesozoic period. Modern forms belong to four orders, distinguished largely by the architecture of the skull.

(a) *Order Rhynchocephalia.*
     This is a primitive group represented today solely by the rare 'living fossil' *Sphenodon punctatus*, the **tuatara** from New Zealand. It has primitive skeletal features and may resemble forms ancestral to the lizards.

(b) *Order Squamata.*
     The familiar lizards and snakes have scales (and often plates too). The skull architecture shows characteristic fenestrations for each subclass. Members of the subclass **Sauria** (lizards) usually have four legs, although some lizards (e.g. the slow-worm *Anguis fragilis*), the 'worm-lizards' (subclass **Amphisbaenia**) and the snakes (subclass **Serpentes**) have lost their limbs and limb girdles (vestigial limbs are occasionally seen in pythons). The skull may be specialized to allow the swallowing of large prey, and some species are venomous. Snakes lack eardrums. Examples include *Lacerta vivipara* (common lizard); *Vipera beris* (adder).

(c) *Order Chelonia.*
     In turtles and tortoises, the body is enclosed in a shell of bony plates fused to the ribs and the vertebrae. Limb girdles lie within the shell. A horny beak replaces teeth. The organ systems of chelonians are relatively primitive. An example is *Chelonia mydas* (green sea turtle).

(d) *Order Crocodilia.*

Crocodiles, alligators and caimans are the only living survivors of the Archosauria (ruling reptiles) which once included the 'dinosaurs', pterosaurs and other extinct reptiles of the Mesozoic period. The skin of crocodilians is covered with bony plates with horny scales outside. The lungs are more complex than the simple sacs seen in other reptiles, and the ventricle of the heart is almost completely divided to make an effective four-chambered heart. The skull architecture resembles that of the 'dinosaur' groups and birds. An example is *Alligator mississippiensis* (Mississippi alligator).

(7) **Class Aves.** In evolutionary terms, the birds are a very **homogeneous class**, probably derived from saurischian archosaurs (lizard-hipped 'dinosaurs') of the Jurassic. The skeletons of the earliest birds (e.g. *Archaeopteryx lithographica*) show remarkable similarities to coelurosaur 'dinosaurs' such as *Deinonychus* sp.

The reptilian scales have evolved into **feathers** in birds. The forelimb is modified to form a **wing**, bearing primary feathers on the 'hand' for propulsion, secondary feathers on the forearm for lift and contour feathers for aerodynamic streamlining. Down feathers provide insulation to maintain the **homoiothermy** ('warm-bloodedness') of this class.

The skull has a large cranial vault to contain the big brain. The eye orbits are large (birds use vision as their main sense). There are no teeth in modern birds but there is a horny beak whose form reflects the diet of the species.

The lungs are large and efficient; an associated air-sac system permits a unidirectional air-flow in the lungs, with blood vessels arranged in a cross-current fashion to permit extraction of up to 90% of the oxygen from inspired air. The heart has four chambers: there are separate, parallel systemic and pulmonary circulations.

All birds are oviparous: eggs are laid on land, usually in a nest. Parental care of the young is normal.

There are two modern superorders:

(a) *Superorder Palaeognathae.*

The **ratites** are usually flightless. The palate is specialized. There is considerable debate over whether ratites are ancestral and/or more primitive than other living birds. Examples include *Struthio camelus* (ostrich); *Apteryx haastii* (kiwi).

(b) *Superorder Neognathae.*

This superorder includes all other birds. There are about 20 orders, the largest being the Passeriformes (perching birds, e.g. the robin). Examples include *Gallus domesticus* [domestic fowl (chicken)]; and *Corvus corone* (carrion crow).

(8) **Class Mammalia**

The mammals are characterized by the possession of hair (fur): scales (e.g. on a rat's tail) are rare. The air trapped within the fur provides insulation to help to maintain the homoiothermy of this class.

The only bone in the lower jaw is the dentary: this articulates with the squamosal bone of the skull (*Fig. 7*); the redundant articulating bones of the reptile are used as auditory ossicles of the middle ear (the maleus and the incus, together with the stapes, hammer, anvil and stirrup). Unless secondarily lost, teeth are present in the upper and the lower jaws: these show considerable diversity of form, adapted to diet.

The young of mammals are suckled on milk secreted by the mother's mammary glands.

The lungs have alveoli. A tidal flow of air into and out of the lungs is assisted by a muscular diaphragm which separates the thoracic and abdominal cavities. The heart has four chambers: as in birds there are separate pulmonary and systemic circulations.

Mammals evolved in the Triassic from synapsid reptiles. There is a well-documented fossil series which leads from mammal-like reptiles (e.g. cynodonts and ictidosaurs) through to the true mammals. Fossil Mesozoic mammals are rare and are usually known from only jaws and teeth.

There are two subclasses:

(a) *Subclass Prototheria (= Monotremata).*

This subclass, with uncertain affinity to both fossil forms and to other modern mammals, lays eggs and the milk is produced by glands rather different from other mammals' mammary glands. The skeleton is distinctive. There are three living species, the duck-billed platypus and the spiny anteaters, all from Australasia. An example is *Tachyglossus aculeatus* (short-nosed echidna).

(b) *Subclass Theria.*

The Theria ('beasts') includes all other living mammals. There are two infraclasses:

- **Infraclass Metatheria (= Marsupialia).** In marsupials, the young are born very immature; they crawl through the mother's fur to the mother's abdomen where they attach to a nipple, usually in a pouch. Most marsupials are native to Australia where the infraclass has radiated widely, but the opossums are indigenous to the Americas. Marsupials have epipubic bones on the pelvic girdle. An example is *Setonix brachyurus* (quokka).

- **Infraclass Eutheria (= Placentalia).** In placentals, the young grow within the uterus, using a chorioallantoic placenta. The degree of maturity at birth varies (cf. newborn mice and guinea-pigs). The infraclass has radiated widely. About 15 living orders are recognized, including:

  1. Insectivora

  A primitive, arguably ancestral order retaining a vestigial cloaca and having a rather lower body temperature than other placentals. An example is *Sorex araneus* (common shrew).

  2. Chiroptera (bats)

  Forelimbs are extensively modified to form a wing of skin. An example is *Pipistrellus pipistrellus* (common pipistrelle).

  3. Primates

  Large eyes and brains, and with prehensile hands and (often) feet. Opposable thumb and the claws replaced by nails. Otherwise rather unspecialized: includes prosimians (e.g. lemurs and bush-babies), old and new world monkeys, apes and humans. Examples include *Nycticebus coucang* (slow loris); *Macaca mulatta* (rhesus monkey); *Pan troglodytes* (chimpanzee).

  4. Carnivora

  Normally flesh-eaters, usually with four clawed digits on each limb. Examples include *Panthera leo* (lion); *Odobenus rosmarus* (walrus).

5. Cetacea

Whales and dolphins are wholly aquatic; the hind limbs are lost. Some have teeth, others filter plankton with baleen plates on the palate. An example is *Tursiops truncatus* (bottle-nosed dophin).

6. Proboscidea

The elephants have reduced numbers of teeth; the upper incisors are modified to form tusks, and the nose and upper lip are extended to form a trunk. An example is *Loxodonta africana* (African elephant). [Arguably related orders include the **Hyracoidea** (the hyraxes) and the **Sirenia** (the dugongs and manatees: marine herbivores.]

7. Artiodactyla

The even-toed ungulates, usually with two hoofed digits on each limb. They are omnivores or herbivores, including ruminants with compound stomachs adapted to the symbiotic digestion of cellulose. Horns are common. Examples include *Cervus elaphus* (red deer); and *Camelus dromedarius* (Arabian camel).

8. Perissodactyla

The odd-toed ungulates are herbivores with an odd number of hoofed digits on each limb, e.g. one in horses. Examples include *Equus berchelli* (zebra); *Diceros bicornis* (African black rhinoceros).

9. Rodentia

The rodents are a very large and successful order (>3000 species). They have a single pair of upper and lower incisors for gnawing: the teeth grow continuously. There is a wide gap (diastema) between the incisors and grinding molars. An example is *Rattus rattus* (black rat).

10. Lagomorpha

The rabbits, hares and pikas are similar to the rodents but possess two pairs of upper incisors. An example is *Oryctolagus cuniculus* (European rabbit).

There are also three orders of *edentates* (armadilloes, sloths, anteaters, pangolins and the aardvark) which tend to eat ants, termites and other insects: they are characterized by long snouts, long, adhesive tongues and reduced or absent teeth.

B1 BODY PLANS AND BODY CAVITIES

Key Notes

Diploblastic	Made up of two layers: the outer ectoderm and the inner endoderm (e.g. Cnidaria).
Triploblastic	Made up of three layers: the outer ectoderm, middle mesoderm and the inner endoderm (e.g. Platyhelminthes).
Body cavities	*Acoelomate*: no enclosed body cavity (e.g. Platyhelminthes). *Pseudocoel*: the body cavity is derived from the blastcoel and is lined by endoderm on the inside, and mesoderm on the outside (e.g. Nematoda). *Coelom*: the body cavity is lined by mesoderm on inside and outside, formed by splitting of mesodermal cell mass (schizocoely, e.g. Annelida) or from outpocketings of primitive archenteron/gut cavity (enterocoely, e.g. Chordata). *Hemocoel*: the blastocoel persists to form the main body cavity filled with blood (e.g. Arthropoda).
Related topics	Skeletons (B3) Protostomes and deuterostomes (B4)

Diploblastic

Diploblasty (*Fig. 1a*) is characterized by an outer **ectoderm** (epidermis) and an inner **endoderm** (gastrodermis), lining a gut (enteron or gastrocoel). It is characteristic of Cnidaria (sea-anenomes, jellyfishes, etc.). The gut can be closed at the mouth to provide a liquid-filled chamber of constant volume to assist support; however, this support role is compromised when the mouth is opened to ingest food or to eliminate waste.

Triploblastic

Triploblasty (*Figs 1b–e* and 3) is characterized by an outer **ectoderm**, a middle **mesoderm** and an inner **endoderm** (lining the gut). It is characteristic of most Metazoa (apart from Mesozoa, Cnidaria and Ctenophora).

Body cavities

Body cavities permit rapid movements (by predators or by prey): the body cavity can function as a **hydrostatic skeleton** through containing noncompressible but deformable fluid, and so transfer forces generated by muscles to other parts of the body.

Acoelomate

In the **acoelomate** (*Fig. 1b*) condition, there is no enclosed body cavity. Support is effected by the turgor of the tissues contained within the ectoderm (which may be bounded by a tough cuticle). This condition is characteristic of the Platyhelminthes (e.g. flatworms such as liverflukes).

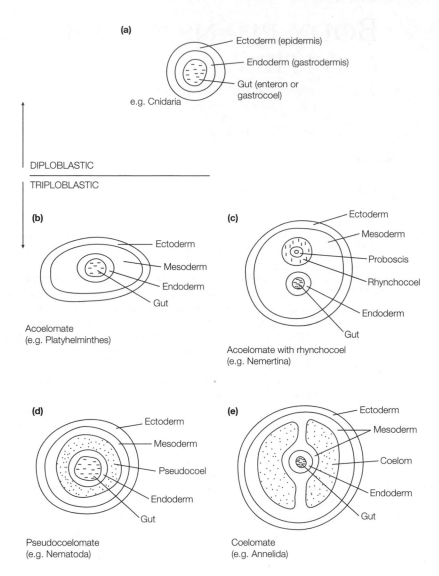

Fig. 1.   Body plans and cavities.

In the acoelomate Nemertina (ribbonworms), a **rhynchocoel** (*Fig. 1c*) surrounds the retracted proboscis. The fluid within this cavity allows muscular
force to be transmitted, resulting in eversion of the proboscis. This localized
phenomenon uses the same principle as is found in a hydrostatic skeleton.

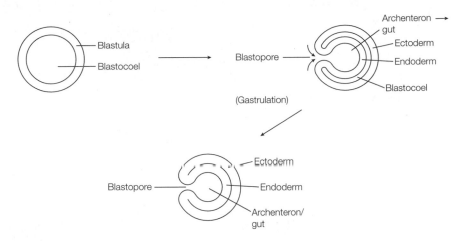

Fig. 2.   Blastocoel reduction during gastrulation.

Pseudocoelomate

In the **pseudocoelomate** (*Fig. 1d*) condition, a body cavity is derived from the **blastocoel** (*Fig. 2*), lined by the endoderm and the mesoderm-derived body wall muscles. It functions as a hydrostatic skeleton and can be used to house organs such as the gonads. It is characteristic of Nematoda (roundworms) and Rotifera.

Coelomate

In the **coelomate** (*Fig. 1e*) condition, there is a **coelom** or 'true' body cavity, lined with muscles (derived from mesoderm) both internally and externally, thus permitting a better control over the movement of the coelomic fluid. It is characteristic of Annelida (e.g. earthworms and ragworms) and Chordata. Further control is facilitated by the separation of the coelom into compartments, as in the transverse **metameric segmentation** seen in annelid worms.

   In some coelomates (protostomes) the coelom forms by the splitting of the mesodermal cell mass (**schizocoely**). In deuterostome coelomates the coelom forms from outpocketings of the **archenteron** (the cavity which results from gastrulation and will give rise to the gut lumen); the pockets separate from the archenteron and fuse to form the coelom, the walls being the mesoderm (**enterocoely**). (See Topic D2.)

Hemocoel

The blastocoel is largely eliminated in the development of forms such as annelids. It persists to form the main body cavity, the **hemocoel** (*Fig. 3*), in Mollusca and particularly Arthropoda where the hemocoel is filled with 'blood' (hemolymph) which bathes the body organs in the open circulation; the true coelom is greatly reduced.

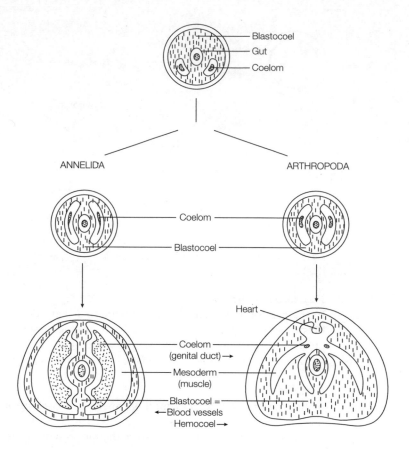

Fig. 3. Hemocoel formation in Arthropoda.

B2 SYMMETRY IN ANIMALS

Key Notes

Radial symmetry	The two halves of the animal are mirror-images of each other, regardless of the plane of cut through the central line
Bilateral symmetry	The right and left halves of the animal, divided into two longitudinal halves, are mirror images of each other.
Asymmetry	The animal is not symmetrical.
Related topic	Body plans and body cavities (B1)

Radial symmetry

Radial symmetry (*Fig. 1*) is found in Cnidaria (e.g. sea-anenomes and jelly-fishes). Two halves of the animal are mirror images of each other, regardless of the plane of the cut (through the center line).

A **secondary radial symmetry** can be superimposed on bilateral symmetry (see below), as in the pentaradiate (five-rayed) symmetry of Echinodermata (e.g. starfishes).

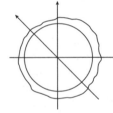

Fig. 1. *Radial symmetry (symmetry with respect to all axes, e.g. Cnidaria).*

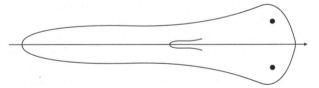

Fig. 2. *Bilateral symmetry (symmetry with respect to one axis, e.g. Platyhelminthes).*

Bilateral symmetry

Bilateral symmetry (*Fig. 2*) describes a condition where only the right and left sides of an animal, divided exactly into two longitudinal halves down the center of the front and back, are mirror images of each other. Most metazoan animals (with the exceptions of Cnidaria and Ctenophora) are bilaterally symmetrical.

Asymmetry

Asymmetry is seen primitively in animals such as sponges (although this asymmetry is arguably imposed on an initial bilateral symmetry). In 'higher' Metazoa, asymmetry can be imposed on bilateral symmetry, as in the architecture of the great arteries leaving the mammalian heart or the differential sizes of the right and left chelae (grasping claws) or a lobster.

B3 SKELETONS

Key Notes

Hydrostatic skeletons	Incompressible but deformable fluid in a cavity of constant volume. Antagonistic circular and longitudinal muscles change the cavity shape. There are large muscle blocks with no leverage (e.g. Annelida).
Exoskeletons	A hard external skeleton, usually waterproof and gasproof. This necessitates the evolution of limbs as levers with discrete, specialized and localized muscles (e.g. Arthopoda – *arthropodization*). Respiratory organs are usually developed.
Endoskeletons	The internal skeleton grows with the animal, giving support and leverage (e.g. Craniata: formed of cartilage subsequently replaced by bone).
Related topic	Body plans and body cavities (B1)

Hydrostatic skeletons

The hydrostatic skeleton works on the principle of an incompressible but deformable fluid contained within an enclosed, flexible cavity with a constant volume (e.g. a pseudocoel or a coelom). The fluid provides support for the body; contraction of muscles changes the shape of the cavity so that forces generated by muscles can be transferred to other parts of the body.

Further control is effected by splitting the cavity transversely into sections or compartments, as in the **metameric segmentation** of an earthworm (an annelid): thus parts of the body can move differentially with respect to others. Shape changes are facilitated by antagonistic muscles (*Fig. 1*), such as circular versus longitudinal muscles.

'Local' hydrostatic skeletons are found, as in the tube-foot of starfishes (*Fig. 2*) or the rhynchocoel and proboscis of nemertine ribbonworms; the principle is also used for penis erection in male mammals. Hydrostatic skeletons are characterized by large muscle blocks and usually no magnification of muscle velocity or power through leverage. They suffer from the disadvantages of relative slowness and problems if the fluid-filled cavity is punctured.

Exoskeletons

Animals with hydrostatic skeletons tend to be rather slow moving and have relatively thin external coverings, permeable to water and to gases (and so prone to desiccation on land and to abrasive damage and predation). An exoskeleton can give mechanical and physiological protection (and also be used for support).

Arthropodization

The development of a rigid exoskeleton protects the body from predators and abrasion, but if the exoskeleton is rendered waterproof to prevent desiccation it will also be gasproof, and organs for respiration (e.g. tracheae) are then required. The exoskeleton, often made from chitin, a strong, flexible

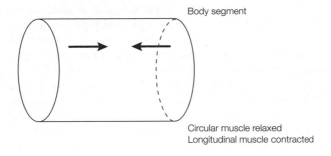

Body segment

Circular muscle relaxed
Longitudinal muscle contracted

Circular muscle contracted
Longitudinal muscle relaxed

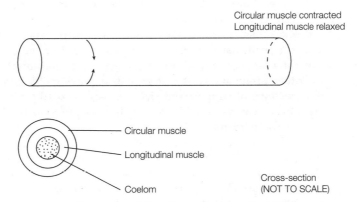

Circular muscle

Longitudinal muscle

Coelom

Cross-section
(NOT TO SCALE)

Fig. 1. Antagonistic muscles of the body wall in an animal with a hydrostatic skeleton, e.g. an earthworm.

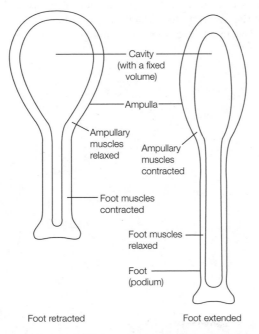

Cavity
(with a fixed
volume)

Ampulla

Ampullary
muscles
relaxed

Ampullary
muscles
contracted

Foot muscles
contracted

Foot muscles
relaxed

Foot
(podium)

Foot retracted

Foot extended

Fig. 2. A hydrostatic skeleton, e.g. tube-foot of starfishes.

polysaccharide, frequently calcified, is also be used for **support**. However, loss of body flexibility associated with development of an exoskeleton (even if components are in sections) favored the evolution of limbs to act as levers for locomotion. Muscles became more discrete and more specialized and localized, seizing on the advantages of levers to promote speed and versatility of movement. Division of labor between body parts was facilitated. The loss of the hydrostatic skeleton function of the coelom resulted in its reduction: the hemocoel containing hemolymph directly bathing the body organs assumed the role of the main body cavity. Arthropoda are the supreme exhibitors of exoskeletons. The possession of a hardened, chitinous, waterproof, gasproof exoskeleton, with the evolution of respiratory organs (such as gills, lung-books or tracheae) and jointed limbs (with specialized muscles), are features of a process which has been termed **arthropodization**.

Exoskeletons do not allow unlimited growth, and large sizes cannot be attained, particularly on land where the weight of the exoskeleton needed to support a large arthropod (e.g. an insect the size of horse) would be excessive. A further disadvantage is that growth requires molting, during which period the problems of support and size are magnified (the new exoskeleton is soft and nonsupportive) and the animal is very prone to predation until the new exoskeleton hardens.

Endoskeletons

An internal endoskeleton is characteristic of Craniata (vertebrates). Here the skeleton can grow within the animal as the animal itself grows, and a larger size is thus possible. The principles of support and leverage seen in exoskeletons also apply for endoskeletons. Protection from desiccation, temperature stress and predation may be given by a waterproof skin which may have scales, feathers, fur or bony plates, although the need for gas exchange will necessitate specialized respiratory organs (e.g. lungs).

The vertebrate endoskeleton appears initially as cartilage, usually replaced by bone comprising hydrated calcium phosphate with protein. (Calcareous endoskeletal elements are found in other phyla such the echinoderms.)

B4 PROTOSTOMES AND DEUTEROSTOMES

Key Notes

Superphyla
Coelomate metazoans can be divided into two 'superphyla': protostomes and deuterostomes.

Protostomes
These have a schizocoelic coelom. They show spiral cleavage of the embryo, with the mouth forming from the blastopore (e.g. Mollusca).

Deuterostomes
These have an enterocoelic coelom. They show radial cleavage of the embryo, with the mouth forming from the second opening opposite the blastopore (e.g. Echinodermata).

Related topics
Relationships between phyla (B5)
Evolutionary origins of the Chordata (B7)

Development and birth (D2)

Superphyla

Coelomate metazoan animals can be divided into two 'superphyla' which are distinguishable on the basis of their early embryonic development. The distinctions may have considerable significance with respect to evolutionary relationships.

Protostomes

Protostomes usually show spiral cleavage of the early embryo (*Fig. 1*), and the coelom forms by schizocoely (see Topics B1 and D2). The mouth arises from the blastopore (*Fig. 2*). Protostome phyla include Annelida, Mollusca and Arthropoda (and arguably phyla such as Nematoda).

Deuterostomes

Deuterostomes show radial cleavage in the early embryo (*Fig. 1*) and they form the coelom by enterocoely (see Topics B1 and D2). The mouth arises from a

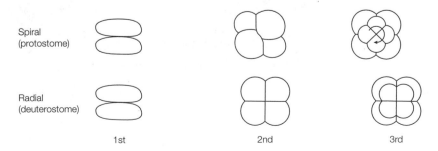

Spiral (protostome)

Radial (deuterostome)

1st   2nd   3rd

Fig. 1. Cleavage of the early embryo, viewed from above.

second opening opposite the blastopore (deuterostome = second mouth) (*Fig. 2*). The blastopore becomes the anus.

Deuterostome phyla include Echinodermata, Hemichordata and Chordata (and arguably the lophophorate phyla). In mammals, the deuterostome pattern of development is no longer obvious.

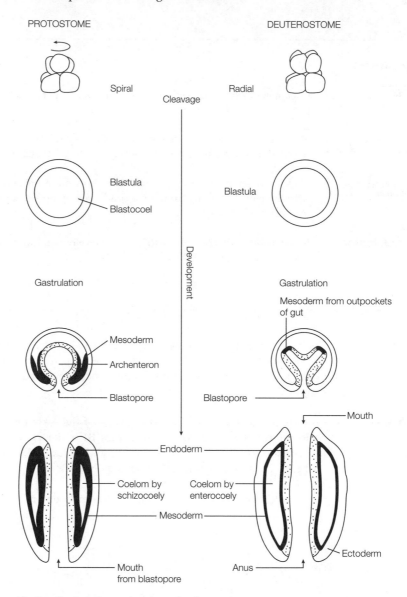

Fig. 2.   Gastrulation and embryo development.

B5 RELATIONSHIPS BETWEEN PHYLA

Key Notes

A possible evolutionary tree can be based on:

- Symmetry
- Diploblastic or triploblastic body organization
- Type of body cavity
- Protostomic or deuterostomic early embryology

Related topics	Body plans and body cavities (B1)	Protostomes and deuterostomes (B4)
	Symmetry in animals (B2)	Evolutionary origins of the Chordata (B7)

There is considerable debate about how animal phyla are related evolutionarily, and many zoologists disagree on the details. Considerations include radial or bilateral symmetry, the former being considered more primitive. There is almost certainly an evolutionary progression from the diploblastic to the triploblastic condition, and from an acoelomate body plan to one with a body cavity, the true coelom being regarded as most advanced. Reduction of the coelom at the expense of a hemocoel (derived from the blastocoel) is also noted.

The protostome group of phyla would seem to form a natural (i.e. evolutionarily related) assemblage, as would the deuterostome group.

Recent studies using molecular techniques such as amino acid sequencing of proteins, nucleotide sequencing of ribosomal RNA and mitochondrial deoxyribonucleic acid (DNA), and DNA hybridization, together with serology (antibody cross-reactivity) tend to confirm the more classical findings of comparative embryology and anatomy.

A possible 'tree' of relationships can be constructed (*Fig. 1*).

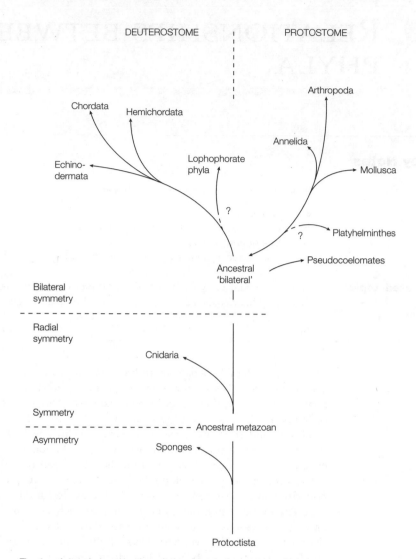

Fig. 1.   A 'tree' showing the relationships between animal phyla.

B6 NEOTENY AND PEDOGENESIS

Key Notes

Neoteny	Neoteny is the retention of larval/embryonic characteristics in the adult (e.g. human skull dimensions).
Pedogenesis	This is the sexual maturity of the larva (e.g. Mexican axolotl).
Related topics	Protostomes and deuterostomes (B4) Metamorphosis (D4) Evolutionary origins of the Chordata (B7)

Neoteny

Neoteny (the retention of larval or embryonic features in the adult) has frequently been involved in animal evolution. It is implicated in theories for the origin of vertebrates. Neoteny has also been put forward as a factor in human evolution, it being suggested that certain features in adult humans (such as dimensions of the skull) resemble those of fetal apes.

Pedogenesis

Pedogenesis (sexual maturity of the larva) also appears to play a role in animal evolution. Many urodele amphibians exhibit pedogenesis: the larva (as in the Mexican axolotl) possesses gills but is sexually mature. Metamorphosis may occur under certain conditions (if the thyroid is stimulated or during drought in the case of the axolotl) or may never be seen to take place in some species (e.g. *Amphiuma means*, the ditch-'snake'). Like neoteny, it is implicated in theories of the origin of the vertebrates.

B7 EVOLUTIONARY ORIGINS OF THE CHORDATA

Key Notes

Comparative embryology	There are similarities between the larvae of echinoderms, hemichordates and urochordates.
Postulated sequence in evolution	● Ancestral arm-feeders (e.g. sea-lilies) move to gill-slit feeding (e.g. acorn-worms). ● Dorsal ciliated bands of echinoderm larvae move into close proximity to form a dorsal nerve tube, facilitated by evolution of the notochord and tail muscles. ● Swimming urochordate larva fails to metamorphose and becomes sexually mature through a process of neoteny and pedogenesis. ● Pedogenetic larva becomes ancestral craniate (= vertebrate); gill-slits change from feeding to respiratory function.
Related topics	Phylum Echinodermata (A12) Protostomes and deuterostomes (B4) Phylum Hemichordata (A13) Neoteny and pedogenesis (B6) Phylum Chordata (A14) Development and birth (D2)

Comparative embryology

The evolutionary origin of the chordates is one of the mysteries of evolution. Like Echinodermata and Hemichordata, the Chordata are deuterostomes, and similarities between the larvae of echinoderms, hemichordates and urochordates (sea-squirts) have long been noted.

It has been suggested that dorsal, longitudinal ciliated bands (*Fig. 1*) in echinoderm larvae moved into close proximity and opposed each other during the course of evolution so that they rolled in and fused to form a dorsal nerve tube: this process would be facilitated by a lateral flattening of the larva caused by the development of a notochord and tail muscles. This theory for the origin of the nerve cord would explain the dorsal position of the sensory columns in the vertebrate spine and the neurenteric canal linking the posterior end of the vertebrate embryo nerve cord to its gut.

Postulated sequence in evolution

An overall pattern of evolution can be postulated, drawing on theories of (among others) Walter Garstang and Alfred Sherwood Romer. The fossil record indicates that ancestral echinoderms were sessile (as are sea-lilies today). Their feathery arms were ciliated and were used to filter-feed particles from sea water. This mode of feeding is shared by lophophorate animals such as bryozoans, where the lophophore is used to filter-feed, and by pterobranch hemichordates such as *Cephalodiscus*.

Thus it seems possible that the chordate ancestor was a sessile arm-feeder; from this ancestor one line gave rise to the primitive and thence the modern

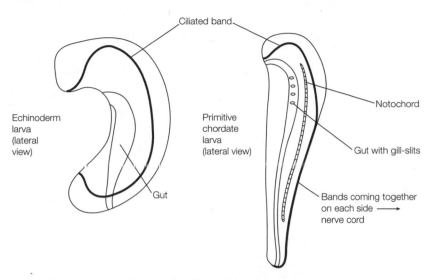

Fig. 1.   Echinoderm and chordate larval resemblances.

echinoderms (starfishes, sea-urchins, etc.), another gave rise to ancestors of pterobranch hemichordates. Within this latter group, there was a shift from arm-feeding to gill-slits: even in some pterobranchs there is a single pair of gill-slits which helps water flow into the gut. As the number and complexity of gill-slits increased, arm-feeding was abandoned. This is seen in the enterop-neust hemichordates (acorn-worms). It is also found, hugely elaborated, in the urochordates (tunicates).

The urochordate larva must now be considered. The echinoderm larva is a poor swimmer, unlike that of the tunicate. The tunicate larva may have devel-oped, with notochord, nerve cord and tail muscles, on the lines suggested above. A powerful, swimming larva is able to seek out suitable positions before meta-morphosis to the sessile tunicate adult.

It is possible that neoteny and pedogenesis were manifested in one group of urochordate larvae. These did not settle but completed their life cycles as swim-ming forms. Such a form could have given rise to an ancestral, filter-feeding vertebrate. The Cephalochordata (amphioxus) appear to be on an evolutionary side-shoot. [That neoteny and pedogenesis occur in urochordates is evidenced by a group of urochordates known as Lárvacea (e.g. *Oikopleura* sp.) where larval features such as a tail are retained into adulthood.] The postulated sequence is illustrated in *Fig. 2*.

The theory can be criticized on the grounds of lack of fossil evidence for many of the intermediate stages. Alternative theories have been suggested: Jefferies invoked a group of stalked, armored, *adult* echinoderms, the Stylophora (e.g. *Cothurnocystis* sp.). These have gill-slits. Progressive modification of this form, including the adoption of a swimming life and loss of armor, could provide a 'ur-chordate' (ancestral chordate): here pedogenesis and neoteny are not implicated.

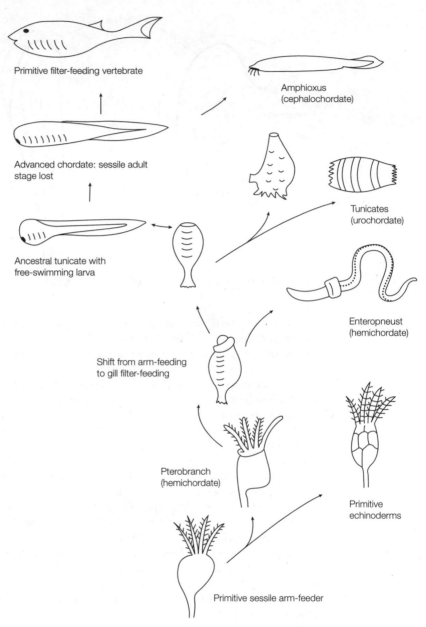

Primitive filter-feeding vertebrate

Advanced chordate: sessile adult
stage lost

Ancestral tunicate with
free-swimming larva

Shift from arm-feeding
to gill filter-feeding

Pterobranch
(hemichordate)

Primitive sessile arm-feeder

Amphioxus
(cephalochordate)

Tunicates
(urochordate)

Enteropneust
(hemichordate)

Primitive
echinoderms

*Fig. 2.   Postulated evolutionary sequence of the Chordata. Redrawn from Romer, A.S.
(1959)* The Vertebrate Story *(4th Edn) with permission from The University of Chicago Press.
© 1933, 1939, 1941 and 1959 by the University of Chicago.*

C1 HOMEOSTASIS

Key Notes

Centrality of homeostasis	Homeostasis is a key concept in the understanding of animal physiology. Health requires constancy of the internal environment in which cells live Homeostasis describes the dynamic equilibium whereby that constancy is maintained through regulation and control.
Variables	Variables within the internal environment include inorganic and organic solute concentrations, pH, hormone levels, body temperature, fluid pressures, etc.
Regulated and controlled variables	Regulated variables are maintained at about the same set-point value: action is taken when there is a departure from the set-point. Controlled variables change to maintain the regulated variable at its set-point value.
System components	Variations in the regulated variable are noted by sensors, compared with the set-point value by interpreters and appropriate action is initiated by effectors; effector responses can be intrinsic (physiological) or extrinsic (behavioral).
Feedback and feedforward	Integral to homeostatic control is negative feedback, a mechanism whereby errors are eliminated through the error triggering processes which correct that error. Positive feedback reinforces error and is useful to initiate a process but, in health, is always overriden by negative feedback. Feedforward anticipates change and initates corrective mechanisms in advance.
Modification of set-points	Set-points can be modifed to meet particular physiological needs.
Related topics	Human blood glucose control (C19) Hormones (C22) Thermoregulation (C21) Integration and control: nerves (C23)

Centrality of homeostasis

The concept of homeostasis is central to an understanding of animal physiology. Claude Bernard, in the mid-nineteenth century, noted that the 'fixité' of the internal environment (*milieu intérieur*) of animals was a condition of 'free life'. For cells within an organism to be healthy and to function efficiently, variables within their internal environment (i.e. the fluids bathing the cells, and the blood and lymph) must be kept within closely defined limits. This relative stability impressed Walter Cannon who, in 1929, coined the term **homeostasis**, noting that it reflected an internal, dynamic equilibrium.

Variables

In the internal environment there are many **variables** which are maintained or controlled at more or less constant levels. These include electrolyte (e.g. calcium) and organic solute (e.g. glucose) concentrations, pH, water balance, pressure, body temperature (in homoiotherms), etc.

Regulated and controlled variables

Two types of variable are recognized. **Regulated variables** stay more or less the same for efficient body functioning. Action is automatically taken when there is a departure from the **set-point** (norm, optimal value), returning the value of that variable to the set-point. An example of a regulated variable is mammalian body temperature. Set-points can be 'real' as in the concentration of an ion, or 'notional' ('as if') where there may be a balance between the amount of a substance ingested or absorbed and the amount excreted.

Controlled variables change as they act in the service of maintaining the constancy of the regulated variable. Examples include the amount of sweating or shivering to defend the regulated variable of body temperature.

System components

Variations from the set-point are noted by a **sensor** (or monitor or error-detector) which sends messages to an **interpreter** (or comparator) which compares the measured index or factor against the set-point value. Output from the interpreter activates **effectors**. For example, blood temperature is sensed by temperature receptors in the periphery and the hypothalamus; hypothalamic interpreters compare the blood temperature with the set-point value; messages are sent to effectors such as sweat glands, hair-erection muscles, smooth muscles in peripheral arterioles, etc. (*Fig. 1*).

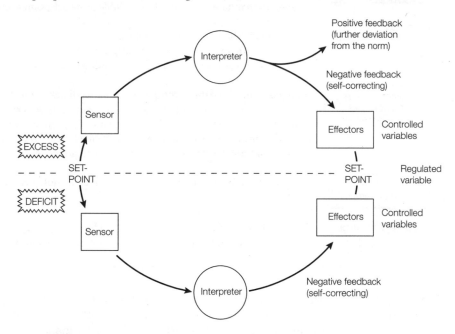

Fig. 1.   A homeostatic feedback system showing system components and feedbacks. The value of the set-point is the regulated variable.

Effector responses can be **intrinsic** ('physiological'), as in sweating or cutaneous (skin) vasoconstriction or vasodilation, or they can be **extrinsic** (behavioral), as in seeking shade, exercising, taking off or putting on clothes.

Variation in heart rate can ensure optimal oxygen delivery to, and levels in, cells: such variation is part of the homeostatic processes of the body, and changes in heart rate can be initiated by low blood pH, high CO_2 levels, etc. The variation in heart rate is said to be **demand-led**.

Variation around the set-point is a part of homeostatic control: the value of the regulated variable oscillates above and below the set-point. Slight excess results in a correction which, through negative feedback (see below), results in a slight deficit which is, in turn, corrected, overshoots, and so on. The amplitude of the oscillations decreases with time, but they are never totally eliminated. This phenomenon is called **hunting about the norm**.

Feedback and feedforward

A key concept in homeostasis is **negative feedback,** whereby an excess or deficit in a system triggers a response which annuls or cancels that excess or deficit: the error is thus **self-eliminating**. For example, excess body temperature triggers vasodilation and sweating; too low a body temperature triggers vasoconstriction, shivering and hair-erection.

Positive feedback would result in reinforcement of the excess or deficit (shivering when too hot!); it does occur, infrequently, in physiological systems, where positive feedback can 'kick-start' a system. Anterior pituitary adenocorticotropic hormone (ACTH) production results in more hypothalamic corticotropin-releasing factor (CRF) release, leading to more ACTH. However, ACTH also leads to release from the adrenal cortex of corticosteroids which have a negative feedback effect on the hypothalamic CRF release. Thus, in health, a positive feedback mechanism will always be overruled later by a negative feedback. Other positive feedbacks include uterine contractions during mammalian birth labor, resulting in more oxytocin release stimulating further contraction, and the sucking reflex whereby the harder the mammalian infant sucks at the nipple, the more milk flows.

Feedforward is also observed and is an important component of many homeostatic control mechanisms. Anticipation of food leads to saliva secretion; food in the mouth or stomach leads to insulin release; rats drink before a meal; humans anticipate the amount of clothing they will need on a hot/cold day; birds migrate to warmer climates in winter while small mammals may hibernate. Note that some of these processes are intrinsic, others extrinsic.

Hemorrhaging in humans illustrates positive and negative feedback (*Fig. 2*). Normally the heart pumps blood at about 5 liters per minute. If 1 liter of blood is lost, pumping effectiveness drops very sharply, but is restored within about

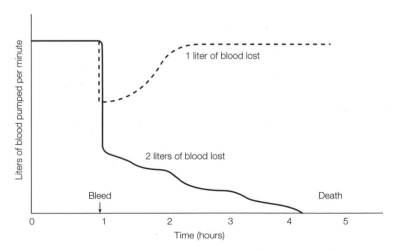

Fig. 2. The effects of hemorrhaging in humans on heart pumping over time.

2 hours. Vasoconstriction mediated by the renin–angiotensin system and arterial smooth muscle contraction, together with the processes of hemostasis, exerts an overall negative feedback to maintain arterial blood pressure, cardiac output and oxygen delivery to tissues.

If 2 liters of blood are lost, the pumping effectiveness of the heart falls over a 2-hour period, leading to death. This suggests that the negative feedback compensation is overridden by other factors having positive feedback elements. (As blood is lost, arterial pressure drops; therefore, blood flow in the coronary circulation to the heart muscles falls, cardiac muscles receive less oxygen, pumping efficiency declines yet further and so on, in a vicious cycle.)

Modification of set-points

Set-points can be modified; for examples in fever in mammals, the temperature is regulated higher than normal to combat infection. Thus a higher temperature (e.g. >39°C) will be defended against challenges. This explains shivering, despite a high temperature during, say, influenza, where pyrogens stimulate 'cold neurons' which initiate heat-generating effector processes.

C2 RESPIRATION

Key Notes

Requirements for gas exchange	Gas exchange needs moist gas-permeable membranes, high oxygen tension in water or air, and a high carbon dioxide tension in the body fluids.
Obtaining oxygen	Gas exchange through the body surface directly to the tissues is only feasible for small organisms. Simple diffusion is possible through a moist external surface to the transport system (blood), for example in the earthworm. Specialized respiratory organs are also used.
Specialized respiratory organs	Respiratory organs include: ● air-ducts direct to the tissues (tracheae in insects); ● evaginations of the body surface (gills in crabs and fishes), where containment in a chamber assists the control of water flow over surfaces; ● invaginations of the body surface (lungs in birds and mammals). Lungs and tracheae reduce the inherent dangers of desiccation through respiratory surfaces in air. Unidirectional flow of water/air and countercurrents in fish gills/bird lungs facilitate efficient oxygen extraction.
Related topics	Human external respiration (C3) Blood and circulation (C7) Gas transport in blood (C4) The mammalian heart (C10)

Requirements for gas exchange

Aerobic metabolism in animal (and in plant) cells needs **oxygen**. The end products, **carbon dioxide** (and water), must be removed from the body. This exchange of gases is called **respiration**. Respiration can be considered at three levels:

● **External respiration** (at the whole organism level): gas exchange between the environment and specialized respiratory organs.
● **Internal respiration**: gas exchange between body fluids and cells.
● **Cellular respiration**: the use of oxygen within the cells.

External respiration will be considered here, at the level of the **whole animal**. One liter of pure water in equilibrium with air at 15°C contains about 7 cm^3 of oxygen; warmer water or water containing dissolved salts (e.g. sea water) will contain less oxygen while colder water will contain more oxygen. This means that at 15°C a unit mass of oxygen is contained within a mass of water approximately 100 000 times as great. Air contains 20.93% oxygen: this means that a unit mass of oxygen is contained within a mass of air approximately 3.5 times as great. Water is also about 1000 times as dense and 100 times as viscous as air. Thus, to obtain a unit mass of oxygen, 10^5 times the mass of water or 3.5 times the mass of air must be moved. Consequently, to conserve energy, water-breathing animals tend to use a **unidirectional** flow of medium across their respiratory surfaces, while **air-breathing animals** can use a **tidal** flow of medium; for example, compare a fish gill with a human lung.

For respiratory **gas exchange** to occur the following are needed:

- a **moist, permeable membrane;**
- body fluids with a **high carbon dioxide tension** (partial pressure) on one side of the membrane;
- air or water with a **high oxygen tension** on the other side of the membrane.

Each gas behaves independently of the other (and of nitrogen, the other major, physiologically inert component in air). When a **difference in diffusion pressure** exists on two sides of a membrane there will be a **tendency to equilibrate pressures**. Thus, oxygen crosses a suitable membrane surface into the body while carbon dioxide simultaneously passes outwards.

In many small animals, gas exchange is direct, from air or water through membranes to tissues, but it is more complex in animals with **large bodies** or **impermeable membranes.**

Obtaining oxygen

Simple diffusion across the body surface to tissues
Gas exchange is via the surface membranes to and from water. In Cnidaria and Porifera (sponges), gases diffuse through epithelial cells and thence to those deeper in the body. Some parasitic Platyhelminthes (e.g. *Taenia* sp.) are bathed in the **body or gut fluids** of their hosts, from which they absorb oxygen and to which they void carbon dioxide.

Simple diffusion is **limited by body size**; only a spherical or thread-like animal with a radius of less than 1 mm can use simple diffusion, given known metabolic rates and the oxygen content of natural water. **Flattening** of the animal can help (e.g. Platyhelminthes) or a **modest metabolic rate** or **ramifying channels** for the passage of water (e.g. some Cnidaria). The method is **inappropriate for air** because the large surface area to volume ratio of such small animals would make the animals very prone to **desiccation** through their moist integuments.

Diffusion to blood
In most higher Metazoa, external respiration is linked to a **blood transport system** which aids the distrubution of the oxygen to the tissues (and the removal of carbon dioxide from them). In oligochaete worms (e.g. *Lumbricus terrestris*), gases diffuse across the **moist body wall** into the blood. Similar **cutaneous respiration** in seen in Amphibia (e.g. *Xenopus laevis, Rana temporaria*) where this process is important in resting animals, particularly during low temperatures in the winter; the **floor of the mouth in frogs** is similarly used. In a few **lungless salamanders**, this is the **sole means** of oxygen uptake. The **disadvantages** of this method include the low rate of delivery of oxygen to the animal (it could not sustain an active, summer frog) and the dangers of desiccation through a surface which can easily become damaged.

Specialized respiratory organs

Tracheae
Tracheae are found in insects, millipedes, centipedes and some arachnids, all of which are members of the **Arthropoda**, and in *Peripatus* (in a rudimentary form). Tracheae are ingrowths of the body wall, lined with chitin; tracheae branch repeatedly and end as microscopic **tracheoles** (without chitin) between

individual cells, which they sometimes indent (*Fig. 1*). The tracheal network forms a 'capillary' network of air-ducts which could be described as a whole-body lung analog. The tracheole tips are filled with fluid through which gases diffuse into and out of adjacent cells. The tracheolar fluid is withdrawn near active cells.

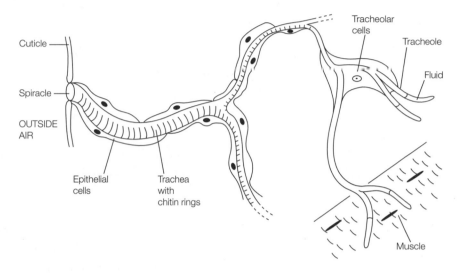

Fig. 1.   The structure of the insect tracheal system.

The **efficiency** of the system reflects the much more **rapid diffusion rate of oxygen in air** ($\approx 43 \times 10^3$ that in water), and the **small size of the arthropods**. This method of transporting oxygen to the cells is very efficient in an animal where gas transport using the blood would be excessively slow because the circulation is open and operates at low pressure.

Tracheae open to the outside by **spiracles** which **may be valved**. In many large insects (e.g. locusts), the spiracles may open and close periodically to reduce water loss; this may be developed into **cyclic respiration** in which there are long periods between spiracle opening during which carbon dioxide builds up to a critical level in the tissues. Some larger insects may assist tracheal diffusion by **muscular pumping of compressible air-sacs** off the tracheae: phased, differential opening of the spiracles may allow a unidirectional flow of air through the larger tracheae.

Aquatic insects may have a **closed tracheal system** in which the openings of the tracheae are covered by a very thin layer of cuticle through which oxygen diffuses from the water. Tracheae may be extended into **'tracheal gills'** to provide a greater diffusive surface (e.g. mayfly larvae); the 'gills' may be housed in the rectum, and water may be pumped in and out of the rectum, to ventilate the gills (e.g. dragonfly larvae).

Gills

Gills and lungs allow gas exchange to take place across expanded surfaces into or from the blood. Gills are evaginations of the body, covered with a thin epithelium. They contain many blood capillaries. Oxygen can diffuse in through the

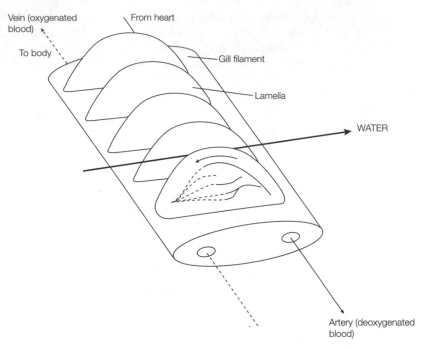

Fig. 2. The structure of fish gills and function of the countercurrent system.

large gill surface area while carbon dioxide diffuses out. **External gills** (e.g. in small crustaceans, in the neotenic axolotl *Ambystoma mexicanum*, or in young anuran tadpoles) are exposed to the water: they can be gently moved, but are easily damaged and present resistance to locomotion. There is little control of ventilation. **Internal gills** (as in decapod crustaceans such as crabs, or in fishes) are contained within **gill chambers** where they are protected and where a controlled flow of water can be used to **ventilate** them.

Opening and shutting the **mouth and opercular cavities** in **teleost** fishes permits a nearly continuous, unidirectional pumping of water across the gills. The gill surfaces are vastly extended on the gill filaments into **lamellae** in which are many capillaries: the blood flow is arranged opposite to the flow of water, so that an efficient **countercurrent system** functions to allow **maximum extraction of oxygen** from the water (about 90% in some species) (*Fig. 2*). **Ram breathers** (e.g. the tuna fish) swim with their mouths open, forcing water across the gills; **cyclostomes** (e.g. the lamprey) cannot take water in through the mouth when they are attached to their hosts so they use **pouch gills** with muscles, permitting a **tidal flow** of water in and out of the gill slits.

Gills are at a **disadvantage in air**, where they tend to collapse or surface tension causes the lamellae to cling together. The **air-breathing teleost** *Periophthalmus* sp. has **stiffened gills** (and it even drowns in water!); some crustacean **woodlice** have **gill-books** for use in air, but they need to moisten them periodically in water.

Lungs

Lungs are **invaginations** of the integument, adapted for respiration. They are found in some arthropods such as **scorpions** (lung books), in **pulmonate** gastropod molluscs (a vascularized mantle cavity) and in **vertebrates** (including several groups of bony fishes). The invagination **prevents desiccation**, but there needs to be a mechanism for **ventilating** the lung.

The internal surface area of the lung may be greatly increased by folds or by pockets or sacs, **alveoli**. In **anuran amphibians** (e.g. frogs), the lung is filled by **pressure pumping** through raising the floor of the closed mouth/nasal cavity; other land vertebrates fill the lung by **suction**, expanding the body cavity through **raising the ribs** (and lowering a **diaphragm** in mammals). In **mammals**, the flow of air is **tidal**; in **birds** the lungs are extended into a series of **non-vascularized air-sacs**. This, together with a cross-current **capillary system** in the lungs, allows a **unidirectional flow of air** through the the lungs, enabling the removal of a much higher proportion of oxygen from the air (up to 90% in birds as opposed to about 25% in mammals): this sustains the high metabolic rates of birds and allows them, for example, to fly at altitudes higher than mammals could tolerate.

C3 HUMAN EXTERNAL RESPIRATION (VENTILATION OR 'BREATHING')

Key Notes

Breathing	Gas exchange occurs in the alveoli of the lungs. The alveoli have a surface area of about 70 m^2.
Respiratory mechanics	Contraction of the diaphragm and intercostal muscles lowers the thoracic cavity pressure and permits inspiration; relaxation of muscles permits expiration. The resting tidal volume of lungs is about 500 ml. Phospholipid surfactants reduce the surface tension in the alveoli.
Control of respiration	Control of the respiratory rate is exercised by the medulla oblongata of the hind-brain. Increased carbon dioxide tension in the blood is the primary signal for an increased ventilatory rate in humans (and other air-breathers: oxygen is the signal in water-breathers).
Related topics	Respiration (C2)
	Gas transport in the blood (C4)

Breathing

In humans, gas exchange takes place within the two **lungs** contained within the **thoracic cavity**. Surrounding the elastic lungs are two thin membranes, the **pleura**, which secrete a lubricating fluid. **Inspiration** (breathing in) and **expiration** (breathing out) usually take place through the nose. The nasal cavities are lined with hairs and cilia which trap dust; the nasal epithelium secretes a mucus which also collects debris and humidifies the air; the cavities have a rich blood supply which keeps the temperature high and warms the air. **Desert mammals** such as jerboas tend to have long nasal passages: these act as countercurrent exchangers to warm and humidify air as it enters, and to cool air and condense (and so conserve) water from the air as it is expired.

Air then passes to the **pharynx** and thence to the **larynx**, a triangular box across which are stretched the two ligamentous **vocal cords**. Vibration of the cords by expired air causes the sounds (duly modified by mouth shape, and lip, teeth and tongue position, and resonated within the pharynx and mouth cavities) which make speech.

The air then passes down the **trachea** (wind-pipe), a long, tube lined with ciliated epithelium and reinforced with cartilaginous C-rings (to prevent its collapse as food is pased down the adjacent esophagus). The trachea branches into the two **bronchi** which subdivide into smaller and smaller air-tracts, the **bronchioles**. These are lined with smooth muscle whose contraction adjusts resistance to air-flow. Cilia lining the tracts waft debris back up to the pharynx.

Gas exchange occurs within **alveoli**, small air-sacs clustered around the ends of the bronchioles. Each alveolus is about 0.1–0.2 mm in diameter and is surrounded by blood capillaries. The alveoli and capillary walls are each one cell thick, separated by thin basement membranes: thus the distance between the air in the alveolus and the blood in the capillary is only about 0.5 μm. Gas is exchanged by diffusion. Human lungs have about 300 million alveoli with a surface area of about 70 m^2, about 40 times the surface area of the body.

Respiratory mechanics

Air flows in and out of the lungs when the air pressure is greater or less than external atmospheric pressure. The pressure in the lungs is varied by changes in the volume of the thorax effected by contraction and relaxation of the **muscular diaphragm** at the base of the thorax (contraction of the diaphragm lowers it and so increases thorax volume); contraction of **intercostal muscles** swings up the rib-cage to increase the thoracic cavity volume further. (In new-born babies, the ribs are a right angles to the vertebral column, so breathing is wholly diaphragmatic.)

Air is forced out of the lungs as the muscles relax. Usually only about 10% of the air in the lungs is exchanged at each breath (the **tidal volume**), but up to 70–80% can be exchanged by deep breathing (*Fig. 1*). The resting tidal volume in most men and women is about 500 ml. When the deepest possible breath is inspired the excess over the resting tidal volume is the **inspiratory reserve volume**: about 1.9 liters in women and 3.3 liters in men. At the end of a forced expiration, the air expelled over the tidal volume is the **expiratory reserve volume**: in women about 0.7 liters, in men about 1.0 liters (*Fig. 1*). The tidal volume and the inspiratory and expiratory reserve volumes represent the **vital capacity** of the lungs. There is a **residual volume** of air left in the lungs (about 1.1 liters in women, 1.2 liters in men) which cannot be expelled (vital capacity + residual volume = **total lung capacity**) (*Fig. 1*).

When the alveoli are empty of air, there is a danger that collapse could occur due to the surface tension of the liquid lining the alveolar surfaces. This is prevented by the presence of a **phospholipid surfactant** (a wetting agent) which

Total lung capacity Men 6.0 liters Women 4.2 liters	Vital capacity Men 4.8 liters Women 3.1 liters	Inspiratory reserve volume Men 3.3 liters Woman 1.9 liters
		Resting tidal volume Men and Women 0.5 liters
		Expiratory reserve volume Men 1.0 liters Women 0.7 liters
	Residual volume Men 1.2 liters Women 1.1 liters	Residual volume

Fig. 1. Average human lung volumes and capacities.

reduces this tension. Lack of surfactant in very premature babies causes major breathing problems: the surfactant appears after about 24 weeks' gestation.

Control of respiration

The rate and depth of respiration are controlled by respiratory neurons in the **respiratory center of the medulla oblongata** (the hind-brain): these are responsible for the rhythmic, automatic processes of normal breathing, and they activate motor neurons in the spine, causing the diaphragm and the intercostal muscles to contract. Periodically the neurons are inhibited, allowing expiration to occur.

Respiratory neurons also receive signals from receptors sensitive to CO_2, oxygen and pH. Chemoreceptors in the bases of the carotid arteries and in the aorta sense lowered oxygen levels in the blood and signal to the respiratory center. The main signal for increased respiratory rate (hyperpnoea) in mammals is, however, **raised blood CO_2** (and lowered blood pH): these are monitored by receptors in the brain-stem itself and by peripheral receptors in the great arteries. CO_2 levels are much more variable in air than are oxygen levels, so CO_2 in the air tends to be the signal for respiratory control in land animals. (In aquatic species, including most fishes, oxygen levels tend to be the signal because the oxygen content in water is much more variable.)

Receptors can detect distension of the lungs – this distension inhibits nervous stimulation of inspiratory breathing muscle contraction (the **Hering–Brauer reflex**). The control system is very sensitive to increased carbon dioxide concentrations [pCO_2] and elevated proton [H^+] concentrations: deliberate hyperventilation leads to dizziness because excessive CO_2 is cleared from the blood (leading to alkalosis) and this inhibits the respiratory center by removing the CO_2 stimulus.

C4 GAS TRANSPORT IN THE BLOOD

Key Notes

Respiratory pigments	Respiratory pigments bind oxygen reversibly.
Vertebrate hemoglobin	The red hemoglobin (Hb) of jawed vertebrates has four globin chains, each linked to one iron-containing heme.
Oxygen–hemoglobin dissociation curve	The oxygen–hemoglobin dissociation curve has a sigmoid profile.
Bohr shift	Acid conditions (such as during periods of increased metabolic rate) shift the dissociation curve to the right, facilitating oxygen unloading at metabolizing tissues (Bohr shift).
Fetal hemoglobin	Human fetal hemoglobin (HbF) has a dissocation curve to the left of that of adult hemoglobin (HbA). This facilitates transfer of oxygen from mother to fetus at the placenta.
Carbon dioxide transport	Carbon dioxide is carried dissolved in plasma (7%), bound reversibly to hemoglobin (23%), and as bicarbonate (70%). The water–carbon dioxide reaction is facilitated by carbonic anhydrase.
Myoglobin	Myoglobin (one globin + one heme) is a storage pigment in vertebrate muscle.
Invertebrate respiratory pigments	Invertebrate hemoglobins are often very large molecules in solution.
Hemocyanin	Copper-based hemocyanin is an important blue respiratory pigment in advanced molluscs and arthropods. It shows a sigmoid dissociation curve and also a Bohr shift.

Related topics	Respiration (C2)	High altitude respiration (C6)
	Human external respiration (C3)	Blood and circulation (C7)
	Diving physiology (C5)	

Respiratory pigments

Oxygen is relatively insoluble in plasma (the liquid part of vertebrate blood): only 0.3 ml of oxygen per 100 ml plasma. Special chromoproteins known as **respiratory pigments** reversibly bind oxygen and transport it.

Vertebrate hemoglobin

In vertebrates, the pigment is red and is called hemoglobin (Hb). Human Hb is made up of four globin protein chains, each linked to an iron-containing heme. The iron in each heme can bind to one oxygen molecule: thus each Hb molecule can carry four oxygen molecules.

Oxygen–
hemoglobin
dissociation
curves

Combination of the first heme with oxygen increases the affinity of a second heme for oxygen, and so on. If a curve is plotted showing the percentage oxygen saturation for Hb against different oxygen tensions (or partial pressures: i.e. oxygen content in the blood), a sigmoid **oxygen–hemoglobin dissociation curve** (*Fig. 1*) is obtained: the S-shape is due to the progressive increase in the oxygen-binding affinity of Hb. Fully oxygenated Hb allows the body to carry 70 times as much oxygen as would the plasma alone. Whether oxygen is taken up or released depends on the oxygen tension (pO_2) of the surrounding plasma: the oxygen tension of the alveolar capillary plasma is high, so Hb tends to load oxygen in the alveolar capillaries; in the tissues it is low so Hb unloads. In tissue capillaries, the oxygen tension is such that Hb does not fully unload and oxygen can be held in reserve for spurts of activity.

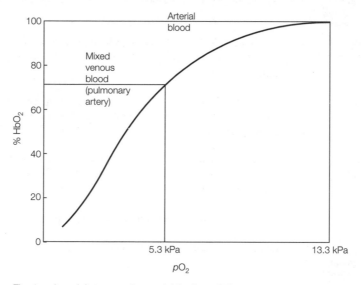

Fig. 1.   An adult oxygen–hemoglobin dissociation curve.

Bohr shift

The oxygen–hemoglobin dissociation curve is shifted to the right by acidity in the blood (this is the **Bohr shift**: *Fig. 2*). Thus oxygen is given up more readily in the tissues, an advantage when muscular work leads to a build up of acid metabolites (including lactate during anaerobic metabolism) and thus there is a need for more oxygen.

Fetal hemoglobin

The fetus has a different hemoglobin (HbF) which has a higher affinity for oxygen (*Fig. 3*) than has adult hemoglobin (mainly HbA in humans with about 2.5% HbA_2 – the different hemoglobins possess different globin chains). Thus the HbF curve is to the left of the HbA curve: this helps oxygen transfer from the mother to her fetus. (Similar polymorphic hemoglobins are found in other vertebrates, e.g. tadpole and adult hemoglobins in frogs.)

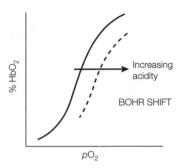

Fig. 2.   Acidity effects on the oxygen–hemoglobin dissociation curve – the Bohr shift.

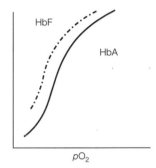

Fig. 3.   The oxygen–hemoglobin dissociation curves for adult hemoglobin (HbA) and fetal hemoglobin (HbF).

Carbon dioxide transport

Carbon dioxide is more soluble than oxygen in plasma, and some 7% of it is simply transported in solution; about 23% of the CO_2 is bound to the globins of Hb. Most CO_2 is carried in the blood combined with water to form bicarbonate (HCO_3^-) ions. This is promoted by the enzyme **carbonic anhydrase** in the red cells:

$$CO_2 + H_2O \underset{\text{Carbonic anhydrase}}{\rightleftharpoons} \underset{\text{Carbonic acid}}{H_2CO_3} \rightleftharpoons \underset{\text{Bicarbonate}}{HCO_3^-} + \underset{\text{Proton}}{H^+}$$

Carbonic acid is formed first; this weakly dissociates into bicarbonate ions and protons. Protons (hydrogen ions) are buffered by combination with deoxygenated hemoglobin:

$$H^+ + \underset{\text{Hemoglobin}}{Hb} \rightleftharpoons HHb$$

The increase in bicarbonate ions facilitates their diffusion into plasma: as they are lost from the red blood cells, the further dissociation of carbonic acid is promoted. More carbonic acid is formed from CO_2 and this favors further uptake of the waste gas from the tissues (*Fig. 4*). [Chloride ions flow into the red cells to correct the removal of negatively charged bicarbonate ions (the chloride shift).] At the lungs these processes are reversed. Note the differences between oxygen and carbon dioxide transport.

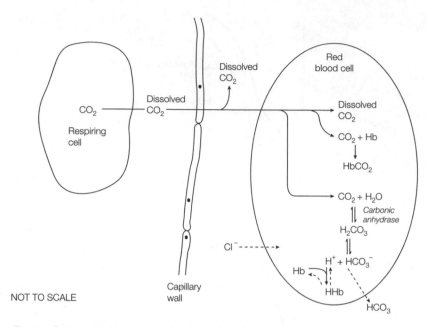

Fig. 4.   Carbon dioxide transport in mammalian blood.

Myoglobin

This red pigment is found in vertebrate skeletal muscle; it has one globin chain with one heme unit. Its rectangular dissociation curve (*Fig. 5*) is characteristic of a storage pigment, and the high oxygen affinity means that it will pick up oxygen from Hb and only loses it at very low oxygen tensions.

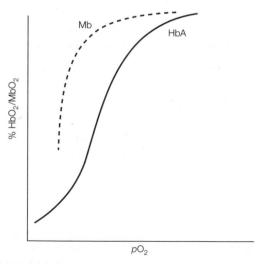

Fig. 5.   Adult hemoglobin (HbA) and myoglobin (Mb) oxygen–hemoglobin dissociation curves.

Invertebrate respiratory pigments

The heme moiety [consisting of four pyrrole rings joined by methane groups to form a super-ring with the ferric ion (Fe^{3+}) at the center attached to the pyrrole nitrogens] is constant for all Hbs, although the globins vary. Hbs seem to have evolved independently many times (they are found in bacteria): a possible mechanism may be from cytochromes – the presence of cytochrome oxidase shows that all animals can combine metalloporphyrins with oxygen. However, there is no oxidation of the ferrous ion (Fe^{2+}) to Fe^{3+}, nor are enzymes involved in the formation of oxyhemoglobin.

Invertebrate Hb-type molecules are often termed **erythrocruorins**: some are enormous. In the polychaete annelid lugworm *Arenicola* sp., the molecule has 180 subunits (unlike the tetrameric vertebrate Hbs) and is carried in solution in the plasma rather than in red blood cells; on the other hand, the polychaete *Glycera* sp. has a dimeric erythrocruorin.

Chlorocruorin

This green metalloporphyrin is allied to Hb and is found in four families of polychaete annelids. In *Serpula* sp., both Hb and chlorocruorin are present, the former being more prevalent in younger worms; in *Potamilla* sp., the Hb is in the muscles and the chlorocruorin in the blood.

Hemerythrin

This nonheme, purple–red, iron-containing respiratory pigment is found in a number of invertebrate groups, including the brachiopod lampshell *Lingula* sp. It is usually in solution in the blood. In *Golfingia* sp., two polymorphs (variants) of the molecule exist. Sixteen atoms of iron combine with eight oxygen molecules, and iron is oxidized from the ferrous to the ferric form.

Hemocyanin

This important blue–gray respiratory pigment is a nonheme, copper-based protein. It is found in molluscs and arthropods, particularly in the more advanced members of the phyla. The whelk *Buccinium* sp. has Hb in the muscles and hemocyanin in the blood. Units of hemocyanin have one copper atom with an approximately 200 amino acid protein chain attached. Two copper atoms are needed per oxygen molecule so a dimer is always present, and some hemocyanins are giant multimers; for example, the snail *Helix* sp. has a hemocyanin with a molecular weight of approximately 6.65×10^6. In deoxyhemocyanin, cuprous copper is always present; oxidation of hemocyanin results in cupric copper. The oxygen affinity of hemocyanin is less than for Hb, but a Bohr shift is demonstrable under acid conditions.

C5 DIVING PHYSIOLOGY

Key Notes

Pressure under water	Pressure increases by approximately 100 kPa (1 atmosphere) per 10 m depth.
Problems for air-breathers	Diving air-breathers face risks of: • aeroembolisms (nitrogen dissolving in blood at depth and bubbling off on resurfacing). Seals and whales have small lungs and exhale before they dive, human divers can replace nitrogen with helium and decompress slowly; • oxygen toxicity (oxygen at high partial pressures is toxic after long periods). Human divers can reduce the percentage of oxygen in the gas mixture breathed; • gas narcosis (nitrogen or helium causes narcosis when breathed at several atmospheres (which limits the duration/depth of dive for humans); • shortage of oxygen.
Oxygen supplies	'Natural' air-breathing divers (e.g. whales, seals, penguins) satisfy oxygen needs when diving by having high levels of hemoglobin and myoglobin, and by possession of large blood volumes per unit body weight.
Natural dives	'Natural' (voluntary) dives are aerobic. The body is charged up with oxygen before the dive by rapid ventilation and a brief period of tachycardia (rapid heart rate).
Forced dives	'Forced' dives result in bradycardia (slowed heart rate), compensatory vasoconstriction diverting blood away from the gut, etc., and anaerobic muscle metabolism.
Related topics	Respiration (C2) Gas transport in the blood (C4) Human external respiration (C3) Blood and circulation (C7)

Pressure under water

Pressure increases by approximately 1 atmosphere ($\approx$100 kPa) for each 10 m depth; the partial pressure of the gases dissolved in the sea reflects their proportions in air (assuming 20% oxygen, 80% nitrogen); therefore, at sea level:

$$pO_2 = 0.2. \text{ atmos } (20 \text{ kPa}), \quad pN_2 = 0.8 \text{ atmos } (80 \text{ kPa})$$

Thus, at 30 m depth, the overall pressure will be about 4 atmospheres (400 kPa) and the partial pressures will be about 0.8 and 3.2 atmospheres (80 and 320 kPa) for oxygen and nitrogen respectively.

Problems for air-breathers

Diving mammals (including humans who scuba-dive) and birds have to cope with the following respiratory-related problems:

- Aeroembolisms
- Oxygen toxicity
- Gas narcosis
- Oxygen shortage

Aeroembolisms (the **bends**) can occur after prolonged diving at depth. A human diving at 30 m must breathe air at 4 atmospheres (400 kPa) pressure (to avoid thoracic collapse due to the hydrostatic pressure of the surrounding water) – **nitrogen will therefore dissolve in the blood** at $0.8 \times 4 = 3.2$ atmospheres (320 kPa). On surfacing to a pN_2 of 0.8 atmospheres (80 kPa), the nitrogen bubbles off in the capillaries causing pain in the joints and dangers of blood flow occlusion, especially in the capillaries of the brain and coronary circulation. Human divers can prevent some of the effects of the bends by breathing a helium–oxygen mixture [helium is less soluble in blood (although it dissolves more quickly)], by making shorter dives, by ascending very slowly or 'decompressing' slowly in a special pressurized decompression chamber. Oxygen is metabolized too rapidly to cause aeroembolisms.

Diving mammals (e.g. whales) and birds (e.g. penguins) avoid aeroembolisms through their possession of relatively small lungs; they expire (breathe out) before a dive. The hydrostatic pressure of the surrounding water forces the residual air in the lungs into the nonvascularized trachea, and so nitrogen under pressure is not taken up by the blood.

High oxygen tensions are toxic and lead to nervous irritation and irritability of the respiratory tract [e.g. oxygen at 3 atmospheres (300 kPa) when breathing air at 15 atmospheres (1500 kPa) at 140 m depth is safe for only about 3 hours]. This is important for human divers. The problem can be ameliorated by breathing a gas mixture with a lower proportion of oxygen (e.g. 95% nitrogen (or helium)/5% oxygen). However, this can lead to gas narcosis.

Gas narcosis can be caused by breathing nitrogen (or helium) at several atmospheres. The result is that human divers who wish to descend regularly to depths of below about 120 m are advised to use rigid devices such as submersibles in which air can be breathed at atmospheric pressure.

Oxygen supplies

Naturally diving mammals and birds need to obtain **oxygen for metabolism** during their dives (unless they metabolize anaerobically). Many divers are able to store relatively high amounts of oxygen: a comparison between a 45 kg seal and a 70 kg man is given below.

	Seal	Man
Lungs	75 ml	800 ml
Blood	1500 ml (25 ml 100 ml^{-1})	1000 ml (20 ml 100 ml^{-1})
Muscle	405 ml (4.5 ml 100 ml^{-1})	240 ml (1.5 ml 100 ml^{-1})
Tissue fluid	150 ml (5.0 ml 1000 ml^{-1})	200 ml (5.0 ml 1000 ml^{-1})
Total	50 ml kg^{-1}	32 ml kg^{-1}

Note that there is a relatively small amount of oxygen in the lungs, but the blood volume is high relative to the body weight. Seal blood has a high oxygen capacity compared with humans (in whales it can even exceed these figures).

The limit is the number of red blood cells per unit volume of blood attainable without making the viscosity of the blood too high. Oxygen is also stored in the muscles.

Natural dives

During a **'natural dive'** (e.g. if a seal dives for fish), the oxygen stores are adequate to maintain the diver in a situation of aerobic metabolism: before an anticipated dive, the diver charges up oxygen supplies by rapid ventilation and a brief period of **tachycardia** (rapid heart rate). Tachycardia and rapid breathing will follow the dive.

Forced dives

In **'forced' or 'escape' dives** (e.g. if a seal is prevented from surfacing), heart rate will fall profoundly (**bradycardia**): this is under immediate nervous control, probably as the result of a cessation of breathing. Arterial pressure is maintained by compensatory vasoconstriction, whereby organs such as the gut receive little blood so that supplies to the brain, heart and eye are maintained. In seals, returning blood accumulates in modified veins such as an expanding vena cava. Muscles work anaerobically. Following surfacing, there is rapid breathing and tachycardia to clear lactate from the system and to replenish oxygen stores.

C6 HIGH ALTITUDE RESPIRATION

Key Notes

Pressure at high altitude

Reduced partial pressure of oxygen at high altitude leads to less 'driving force' to push oxygen into the body.

Initial responses

On ascent, hyperventilation increases arterial oxygen levels and reduces blood carbon dioxide, causing temporary respiratory alkalosis. Alkalosis can inhibit respiratory centers, resulting in cyanosis, hypoxia and impaired mental performance.

Acclimatization

Acclimatization includes increased cardiac rate, release of red cells from spleen leading to polycythemia and erythropoietin-directed erythropoiesis. The oxygen–hemoglobin dissociation curve shifts to the right. There is also an increased capillary density and mitochondrial numbers.

Related topics

Respiration (C2)
Human external respiration (C3)
Gas transport in the blood (C4)

Blood and circulation (C7)
Mammalian blood (C9)
The mammalian heart (C10)

Pressure at high altitude

At sea level, there is approximately 0.2 atmospheres (20kPa) of pressure driving oxygen from the atmosphere to the mitochondria where it is metabolized; above 8000 m this pO_2 is reduced to 0.07 atmospheres (7 kPa). Thus the 'driving force' for oxygen is reduced. Humans can live permanently at 5000 m and can ascend briefly to above 8000 m (e.g. Mount Everest, 8900 m, has been climbed without pressurized oxygen apparatus). Humans can acclimatize to high altitudes.

Initial responses

On exposure to high altitude, lowlanders **hyperventilate** (up to 50% above the sea-level rate over the first 7 days of exposure). Later, there is some compensation (also seen in native Andean highlanders) through deeper breathing. Hyperventilation increases arterial oxygen levels but also reduces blood carbon dioxide levels and raises the proportion of oxygen in the alveolar gas mixture (since the alveolar gases are more or less in equilibrium with the pulmonary blood gases). Thus the pO_2 gradient between inspired air and the gas mixture in the alveoli is reduced, and a sufficiently high pO_2 level is maintained in the alveoli to allow oxygen diffusion into the blood. The reduction in blood pCO_2 causes respiratory alkalosis, corrected by increased kidney excretion of HCO_3^-. Before this compensation occurs, alkalosis can inhibit the respiratory center of the brain leading to hypoxia (shortage of oxygen), characterized by cyanosis (blueing of the extremities such as the lips and ear lobes due to deoxygenated arterial blood) and impaired mental performance, such as judgement and concentration. The cardiac rate increases transiently to improve oxygen transport to the tissues.

Acclimatization   In time, red blood cells are released from reserves in the spleen (leading to **polycythemia**), and new red cells are produced under the influence of the hormone **erythropoietin**. Thus the oxygen content per unit volume of blood in a highlander can be much the same as in a lowlander. Hyperventilation tends to be maintained (highlanders frequently develop 'barrel chests') because of the increased sensitivity of central chemoreceptors to carbon dioxide.

The **oxygen–hemoglobin dissociation curve** in acclimatized lowlanders tends to shift to the right in the tissues because of the build-up of 2,3-diphospho-glycerate, a glycolysis intermediate, and an exaggerated Bohr shift is seen. This assists oxygen unloading at the tissues. Interestingly, in some peoples, such as Tibetans, living at very high altitudes, and in high-altitude artiodactyl species such as yaks, llamas and vicuñas, the dissociation curve shifts to the *left*: this is probably an evolutionary adaptation (rather than acclimatization) and assists loading of oxygen at the lungs.

Increased capillary density and mitochondrial numbers are seen, as are elevated myoglobin levels.

C7 BLOOD AND CIRCULATION

Key Notes

Functions of blood	Blood transports gases, nutrients, wastes, metabolites, defense and messenger substances, cells and heat, and transmits force.
Components of the circulatory system	Blood is contained within a circulatory system comprising vessels, fluid and a pump (heart).
Open and closed circulations	Open circulations (e.g. as in insects) typically have low, unsustained pressure, little flow regulation and a slow return to the heart. Closed circulations (e.g. as in mammals) typically show high, sustained pressure, good flow regulation and a rapid return to the heart.
Invertebrate circulations and hearts	Invertebrate circulations are absent or poorly developed in many phyla, but are well developed in annelids, molluscs and arthropods. Invertebrate hearts are usually dorsal. The insect heart is neurogenic and tubular, and works on a suction pump principle. The vertebrate heart is ventral, compact and myogenic, working on a force pump principle.
Related topics	Gas transport in the blood (C4) Mammalian blood (C9)
	Mammalian circulatory system (C8) Lymphatic system and lymph (C15)

Functions of blood

Except in very small animals (e.g. flatworms) or those with very low metabolic rates (e.g. jellyfishes), simple diffusion from the surface of the body cannot distribute gases and nutrients, metabolites and excretory products, and hormones and defense components to the relevant cells. Most animals contain a transport fluid (blood or hemolymph): localized stirring of the fluid by bodily movement is enough to facilitate transport in simple animals; some form of mechanical pumping must be used in larger, more complex forms.

Blood has the following functions:

- transport of nutrients, metabolites, waste products, hormones and other messenger substances, defense components, heat;
- transport of cells (e.g. thrombocytes, lymphocytes);
- force transmission (e.g. hydrostatic skeleton in oligochaetes, crustacean molt, mammalian penis erection).

The circulatory principle is first encountered in the cytoplasmic streaming of protoctistans.

Components of the circulatory system

In Metazoa, the circulatory system has three components:

- vessels (arteries, capillaries, veins and lymphatics);
- pump (heart);
- fluid (blood, lymph or hemolymph).

Open and closed circulatory systems

Open circulations (e.g. crustaceans, insects, lampreys)

- Low pressure
- No sustained pressure
- Little regulation of blood flow to organs
- Slow return of blood to pump

Closed circulations (e.g. earthworms, cephalopods, most vertebrates)

- High pressure with elastic vessel walls sustaining the pressure
- Regulation of distribution of blood flow to the organs
- Rapid return of blood or hemolymph to the heart

Invertebrate circulations and hearts

In many invertebrate phyla, the coelomic fluid acts as a circulatory system. A circulation is lacking in sponges, cnidarians and flatworms, and is only slightly developed in ribbonworms and most 'protochordates' (e.g. acorn-worms and tunicates).

Invertebrate hearts are very varied in form, but are usually dorsally situated. The **mollusc** heart is dorsal with one or two receptor atria and a single, muscular ventricle pumping blood to arteries leading to organs. In **earthworms**, there are dorsal and ventral longitudinal blood vessels, connected by paired circular vessels in most segments. Pumping is by contraction of the mid-dorsal vessel and five pairs of anterior, lateral 'hearts' connecting the dorsal and ventral vessels.

Arthropod hearts are dorsal and contraction is **neurogenic**. (The rhythmicity of contraction is under the control of motor nerves innervating the heart, unlike the situation in the **myogenic** mammalian heart where the organ has its own intrinsic rhythmicity originating in the pacemaker.) **Crustacean hearts** are compact and roughly spherical; **insect hearts** are slender, elongated tubes (*Fig. 1*) which work on a suction pump principle. Aliary muscles contract to expand the heart volume, and hemolymph (blood) is sucked in through segmental, lateral ostia. Intrinsic muscles in the heart wall contract to propel blood forwards through a dorsal aorta to the hemocoel cavity in which the organs are bathed in hemolymph: valves prevent the outflow of hemolymph through the lateral ostia. Pressure is lost as the hemolymph is passed out into the hemocoel. Insect hemolymph carries food and waste products, etc., but respiratory gases are exchanged through the tracheal system.

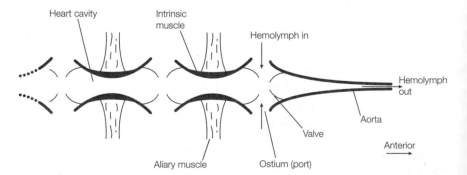

Fig. 1.   Schematic diagram of the anterior part of an insect heart (dorsal view).

C8 MAMMALIAN CIRCULATORY SYSTEM

Key Notes

Arteries	Arteries duct blood from the heart. The arteries have tough, muscular, elastic walls to maintain sustained pressure.
Capillaries	Capillaries have single-cell walls which facilitate exchange in the tissues.
Veins	Veins have thin walls, and blood at low pressure. Valves prevent back-flow.
Circulatory patterns	A typical adult mammal circulatory pattern is: left heart→body→right heart→lungs→left heart. A typical fetal mammal circulatory pattern is: umbilical vein→right heart (the ductus venosus bypasses the liver); the foramen ovale links the right and left atria; the ductus arteriosus bypasses the lungs so that blood flows from the pulmonary arteries to the aorta.
Blood pressure	Blood pressure is related to the strength and rate of heart contraction, the elasticity of vessel walls and the rate of blood flow. The rate is proportional to the radius of arterioles. Arteriolar muscles control the flow to the capillaries by reducing the radius of the vessels.
Related topics	Blood and circulation (C7) The immune system (C12) Mammalian blood (C9) Immune responses (C13) The mammalian heart (C10) Lymphatic system and lymph (C15)

Arteries

Arteries are efferent vessels which carry blood away from the heart to the capillary beds. In adults, all arteries except the pulmonary arteries carry oxygenated blood: the pulmonary arteries carry deoxygenated blood from the heart to the lungs. The great arteries emerging from the heart are the aorta from the left ventricle (carrying blood to the body in the systemic circulation) and the pulmonary trunk from the right ventricle.

Arteries are made up of the following.

- **Tunica intima**: the innermost, smooth, lining endothelium; connective tissue; elastic tissue.
- **Tunica media**: thick connective and elastic tissue and smooth muscle (more muscle in arterioles, more connective tissue in great arteries).
- **Tunica adventitia**: collagen fibers and smooth muscles, with nerves, lymphatics and small blood vessels.

The elastic walls of the great arteries permit adjustment to great pressure during systole (ventricular contraction phase); pulmonary arteries are oval in cross-section, helping to damp surges of blood from the right ventricle and so protect the capillary beds of the lung alveoli.

Arteries branch to form **arterioles**, with extensive smooth muscles supplied with autonomic nerve fibers: thus, they can constrict or dilate and act as 'stop-cocks' controlling blood flow to capillary beds (by $\leq 400\%$).

Capillaries

Capillaries are found at the ends of the finest arterioles and connect the arterial and venous systems. They have walls of only a single layer of endothelial cells. The total length of an adult man's capilllaries is estimated at 96 000 km, giving an enormous surface area for gas, food and waste exchange.

There are three types of capillary (*Fig. 1*):

- continuous capillaries in muscle cells;
- fenestrated capillaries (with pores between adjacent endothelial cells) in kidneys and endocrine glands;
- sinusoids (discontinuous capillaries) with membrane-covered gaps between endothelial cells in liver and spleen.

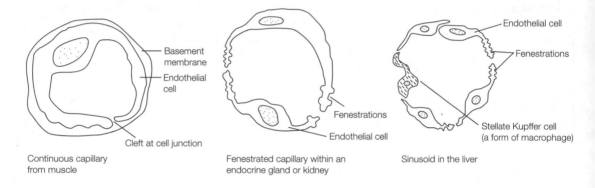

Fig. 1.   The three types of capillary (basement membrane shown in continuous capillary only).

Capillary blood flow
Blood leaves the heart flowing at 30–40 cm sec^{-1} but slows to 2.5 cm sec^{-1} in arterioles and to less than 1 mm sec^{-1} in capillaries, where it stays for a few seconds while gas and nutrient interchange occurs: the single-cell endothelium permits small molecules to pass through while retaining large molecules and cells. Hydrostatic arterial blood pressure forces water and small solutes into the tissue fluid spaces at the arteriolar end of the capillary. The colloidal osmotic pressure of the plasma proteins pulls water and small solutes back into the blood by osmosis at the venous end of the capillary: any fluid not resorbed is drained by the lymphatic system (*Fig. 2*).

Veins

Venules drain capillaries, tiny venules uniting to form veins. The smallest venules are important for the pulling back of tissue fluid into the blood circulatory system. The veins drain deoxygenated blood from the venules to the right atrium of the heart (via the great veins, the **venae cavae**). The pulmonary veins take oxygenated blood from the lungs to the left atrium; the hepatic portal vein takes blood from the capillaries of the gut to the capillaries of the liver. Veins have the same tissue layers as arteries, but the **tunica media** is much

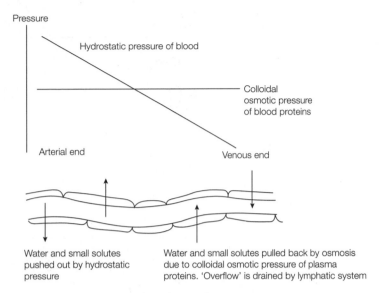

Fig. 2. Fluid flow in and out of capillaries.

thinner and there is less elastic tissue and muscle in the walls: thus veins are very compressible and distensible. They have semi-lunar valves to prevent blood back-flow. Venous pressure is low but flow is assisted by massaging of body muscles in exercise.

Circulatory patterns

Before birth (fetus) the umbilical cord contains the umbilical vein which carries oxygenated blood from the placenta (where interchange of nutrients and gases occurs between maternal blood and fetal blood in capillaries: there is no blood mixing) to the fetal abdomen. Here the vein branches: some blood goes to the liver, most bypasses the liver (via the **ductus venosus**) and joins the inferior vena cava, mixing with deoxygenated blood from fetal tissues.

The vena cava opens into the right atrium. About 67% of the blood passes direct to the left atrium via the **foramen ovale** and hence out via the left ventricle and aorta. Blood from the head discharges into the right atrium: most of this flows to the right ventricle and pulmonary arteries but the **ductus arteriosus** short-circuits the lungs so that blood flows from the pulmonary arteries direct to the aorta; only 10% of the pulmonary artery blood goes to the developing lungs.

At birth, the lungs inflate under adrenaline (epinephrine)/shock stimuli as the baby takes her/his first breath, the pulmonary vessels become fully functional, the pressure on the right side of the heart reduces and the flap over the foramen ovale closes as the right–left heart pressure differential reverses. The ductus arteriosus contracts to form a ligamentous strand; the umbilical vessels atrophy after the cord is severed.

After birth the pattern is as follows: left heart→body→right heart→lungs→left heart.

Figure 3 demonstrates circulatory blood flow patterns.

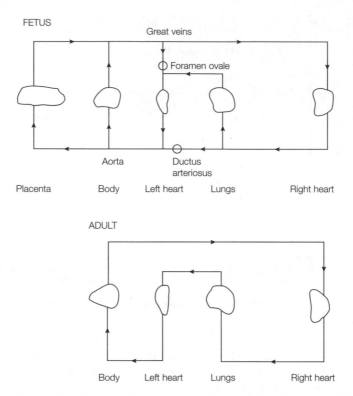

FETUS

Great veins

Foramen ovale

Aorta

Ductus
arteriosus

Placenta   Body   Left heart   Lungs   Right heart

ADULT

Body   Left heart   Lungs   Right heart

Fig. 3.   Simplified blood flow patterns in the mammalian fetus and adult.

Blood pressure Ventricular contraction propels arterial blood with great force: blood pressure measures force per unit area with which blood pushes against vessel walls. Normal adult human arterial blood pressure, measured in the upper arm, is about 120 mm mercury (Hg) when the ventricles contract (**systole**) and 80 mm Hg when they relax (**diastole**): this is written as 120/80. Blood pressure is determined by:

- strength and rate of heart contraction;
- elasticity of arterial walls;
- rate of blood flow.

The rate of blood flow is directly proportional to its pressure. Fluid flow in a tube is proportional to r^4 (where r = radius). The diameter ($2r$) of arterioles can be adjusted by smooth muscle contraction or relaxation, thus profoundly affecting flow rates. Thus the muscles, under the influence of autonomic nerves and chemicals such as adrenaline (epinephrine), exert control.

Arteriolar constriction and dilation regulate flow according to need: flow to muscles increases in exercise, to the gut for digestion and to the skin when hot. A constant flow of blood to the brain is essential (fainting causes a fall so that the force of gravity does not have to be overcome in order to deliver blood to the brain; thus its blood supply is protected). Higher brain activity can affect blood flow: turning pale with fear or anger, blushing with embarrassment, penis/clitoris erection with erotic arousal.

C9 MAMMALIAN BLOOD

Key Notes

Blood composition	Blood is 60% plasma and 40% cells (including thrombocytes).
Plasma	Plasma contains water, plasma proteins (especially albumins, globulins and fibrinogen) and also electrolytes.
Red blood cells	Red blood cells (erythrocytes) are anucleate, biconcave disks, with a life of about 120 days in humans. Erythropoiesis in the adult bone marrow is under the control of erythropoietin. A red cell loses its nucleus and endoplasmic reticulum as it matures.
White blood cells	White blood cells (leukocytes) include granulocytes (phagocytic neutrophils, phagocytic eosinophils, basophils with vasoactive amines involved in anaphylactic hypersensitivity) and agranulocytes (phagocytic/antigen-presenting monocytes/macrophages, and lymphocytes involved in adaptive immunity).
Thrombocytes and hemostasis	Thrombocytes (platelets) and clotting factors are involved in hemostasis; an enzyme cascade amplifies the steps in the clotting reaction. Thromboplastin converts prothrombin to thrombin: thrombin converts fibrinogen to an insoluble plug of fibrin. Control is also exercised by vasoconstriction (autonomic nervous system and renin/angiotensin and other hormones). Anticoagulants prevent unwanted clots; fibrinolysis dissolves redundant clots.
Human blood groups	ABO blood group system: the A group has an A antigen, and an anti-B antibody; the B group has a B antigen, and an anti-A antibody; the O group has no antigen, and both anti-A and anti-B antibodies; and the AB group has both A and B antigens, but neither antibody. These antibodies are immunoglobin (IgM) isohemagglutinins. Rhesus system: the rhesus positive (Rh+) group has a D antigen, which can be a problem if a Rh– mother has a Rh+ baby; the anti-D immunoglobulin (IgG) antibody crosses the placenta at a subsequent pregnancy.
Related topics	Blood and circulation (C7) Mammalian circulatory system (C8) The immune system (C12) Immune responses (C13)	Hypersensitivity, autoimmunity and immunization (C14) Lymphatic system and lymph (C15)

Blood composition A 75 kg human has about 4 liters of blood, comprising about 8% of body weight. About 60% of blood is **plasma**, made up of 90% water, while the remaining 40% of blood is cells and thrombocytes.

Plasma

With the exception of oxygen and carbon dioxide carried by hemoglobin, most food, waste, defense and messenger molecules are carried in solution in the plasma. Plasma also contains **plasma proteins**, the main ones being:

- **albumins** which maintain a high colloidal osmotic pressure in plasma and can bind and carry some hormones: about 60% of plasma proteins are albumins;
- **fibrinogen** (4%) which is made in the liver and is important for blood clotting;
- **globulins** (36%) of which there are three types, α, β and γ. α-Globulins and β-globulins are made in the liver and transport fats and fat-soluble vitamins in the blood: low density lipoprotein transports cholesterol from liver to body cells; high density lipoprotein helps to remove cholesterol from arteries. γ-Globulins are immunoglobulins (antibodies).

Plasma also contains some vitamins, hormones and enzymes (especially concerned with blood clotting) and antimicrobial peptides. **Plasma electrolytes** are inorganic compounds which have ionized in solution. The major cation is Na^+ which has an important effect on fluid movements and helps to determine fluid volume; the main anion is Cl^-. Among other important ions are K^+, Ca^{2+}, Mg^{2+} and PO_4^{3-}.

Red blood cells (erythrocytes)

Red cells make up about 40% of the human blood volume (this is known as the **hematocrit** or **packed cell volume**, strictly including all blood cells). Humans possess 25×10^{12} red cells, each cubic millimeter of blood having 5.5 million red cells (men) or 4.8 million red cells (women).

Red cells are biconcave disks, 7 μm in diameter and 2 μm thick, with a large surface area for diffusion. Mature red cells are made up virtually entirely of hemoglobin: the red cell loses its nucleus and most of its organelles. The thin, strong, flexible, semi-permeable cell membrane facilitates movement through vessels.

Red cell production (erythropoiesis)

In the fetus, red cells are made successively in the yolk-sac, liver and spleen, and (at over 4 months' gestation) in the bone marrow where they are also made in the child and the adult. They start life as stem cells, undergoing a series of divisions during which the nucleus gets smaller and disappears after several divisions: the resulting reticulocyte enters the bloodstream. The extensive endoplasmic reticulum which characterizes the reticulocyte is lost, and the mature erythrocyte, the definitive red cell, is the result.

Adult erythropoiesis makes 10^{10} cells per hour (enough for 440 ml per week, 'one pint'). The process requires amino acids, iron and vitamins (folic acid, riboflavin, B_6 and B_{12}). The controller of erythropoiesis is the glycoprotein hormone erythopoietin, synthesized by kidney cells sensitive to reduced oxygen tensions in the blood.

After about 120 days, red cells become leaky and senescent: they are destroyed by splenic macrophages. Amino acids and iron are recycled, the rest of the heme ends up as bile pigments.

White blood cells (leukocytes)

White cells are divisible into **granulocytes** and **agranulocytes**. White cells are nucleated and can migrate through vessel walls. Their metabolism is more complex than that of red cells and they can synthesize proteins. There are between 4 and 6×10^3 white cells per mm^3, although this can increase by over fourfold following infection or in certain diseases.

Granulocytes

- **Neutrophils ('polymorphs')** are scavenger cells with lobed nuclei which engulf and destroy microorganisms, etc., using an armory of lysosome-derived enzymes. They are the most frequent white cell seen in a blood smear.
- **Eosinophils** phagocytose antigen–antibody complexes and counteract basophil degranulation products.
- **Basophils** contain histamine and other vasoactive amines. When IgE antibody combines with antigen, the basophil (mast cell in tissue) degranulates, releasing the amines and causing inflammation, etc. They are seen in anaphylactic hypersensitivities such as asthma and hay fever.

Granulocytes develop in the bone marrow from stem cells via myeloblasts: about 100 million neutrophils can be produced per day, and have a lifespan about 1 week.

Agranulocytes

- **Monocytes** (mobile macrophages) have large folded nuclei and fine cytoplasmic granules. They originate from stem cells through monoblasts in the bone marrow, and enter the bloodstream for 30–70 hours, before leaving it. In tissue spaces they enlarge to form macrophages, which are important in the liver, lungs and lymphoid organs for ingesting foreign particles in order to destroy them or to present them to lymphocytes.
- **Lymphocytes** are small, nonphagocytic cells with large nuclei. They are concerned principally with adaptive immunity.

Thrombocytes (platelets) and hemostasis

Blood clotting is a complex phenomenon requiring **thrombocytes** (platelets) and at least 15 **factors** present in the bloodstream or on cell membranes. Thrombocytes are colorless, ovoid disks, smaller than red cells, derived from fragmented metamegakaryocytes, derived in turn from bone marrow stem cells; production is under the control of **thrombopoietin**. Platelets are effectively anucleate packets of chemicals needed to initiate clotting of blood and plugging of leaks.

The clotting process begins when plasma encounters a rough surface or a protein **tissue factor** found on the outer surface of many cell types, but not on the endothelial cells lining blood vessels. When tissue factor reacts with a specific circulating plasma protein, for example when a blood vessel is damaged, a cascade of chemical reactions is triggered. In such a cascade, the product of each step of the reaction series acts as a catalyst for the next step, and the molecules involved are, like enzymes, re-used many times. Thus the number of molecules is amplified at each step in the series.

Ultimately, **thromboplastin** is activated. Thromboplastin acts to convert **prothrombin** (a plasma protein produced by the liver) to the enzyme **thrombin**.

$$\text{Prothrombin} \xrightarrow{\text{Thromboplastin}} \text{Thrombin}$$

Thrombin then converts **fibrinogen**, a soluble plasma protein, to **fibrin**.

$$\text{Fibrinogen} \xrightarrow{\text{Thrombin}} \text{Fibrin}$$

The fibrin molecules clump together to form an insoluble mesh which traps blood cells to form a clot. The clot contracts, pulling together the wound edges. (Clotting also occurs if blood is put in a glass tube, apparently triggered by contact with the glass. The clear fluid remaining above the clot is **serum**.)

The autonomic nervous sytem (through its sympathetic fibers) helps to control loss of blood. When about 10% of the blood is lost, the blood pressure suddenly falls and the body enters a state of shock. Reflexes are triggered leading to artery and vein vasoconstriction; the renin/angiotensin system also facilitates vasoconstriction; tachycardia (rapid heart rate) may compensate for reduced blood pressure, delivering extra blood to the brain and heart. This can allow short-term survival if even 40% of the blood volume is lost.

Deficiencies in various clotting factors can lead to failures in the clotting mechanisms: hemophilia is a well-known example: in the most common form there is a congenital absence of Factor VIII which activates prothrombin.

Anticoagulants circulate in the blood to prevent dangerous, unwanted clots: one of the most powerful is the polysaccharide **heparin** produced by basophils. **Clot destruction** is facilitated by **fibrinolysis**, a process by which **plasmin** digests fibrin.

Human blood groups

ABO system

If red blood cells from one person are mixed with plasma from another, **agglutination** of red cells may occur due to isohemagglutinins (IgM antibodies) of the ABO blood group system. The relevant antigens are A and B (derived from a sugar, H substance) on red cells. The antibodies may derive from reactions to dietary sugars or to polysaccharides on the surfaces of gut microflora. There are four blood groups; people in:

- group O have neither A nor B antigen
- group A have A antigen
- group B have B antigen
- group AB have both antigens

- group O have anti-A and anti-B antibody
- group A have anti-B antibody
- group B have anti-A antibody
- group AB have neither antibody

People in group O are known as 'universal donors' as they can give blood to any recipient, and those in group AB are known as 'universal recipients' as they can receive blood from any donor.

Rhesus system

Most people have the **D antigen** on their red cells: they are **Rhesus positive (Rh+)**; the rest are **Rh−**. If a Rh- mother has a Rh+ baby (the gene for D being inherited from the Rh+ father) there are usually no problems with the first pregnancy, but mother will probably be immunized against baby's D antigens because of bleeding at birth. IgG antibodies to D are made: these can cross the placenta. Thus a second or subsequent Rh+ baby is at risk from anti-D antibodies which cross the placenta and destroy the fetus' red cells, leading to **hemolytic disease of the newborn**. Treatment is to give Rh– mothers passive anti-D antibody immediately after birth; this destroys any Rh+ red cells from the baby's blood before they can immunize the mother.

C10 THE MAMMALIAN HEART

Key Notes

Heart structure	The heart has four chambers. The right and left atria act as priming pumps, and the right and left ventricles act as force pumps. The right heart pumps from the body to the lungs, and the left heart pumps from the lungs to the body. Valves prevent back-flow from ventricles to atria, and from great arteries to the ventricles.
Heartbeat	The ventricular filling phase is known as diastole. The ventricular output (pumping) phase is known as systole.
Heartbeat control	Myogenic (intrinsic) rhythm is initiated in the pacemaker (the sinoatrial node). A normal rhythm is about 72 beats per minute. Rapid conduction of impulses occurs through the atria. Impulses reach the atrioventricular node with an approximately 100 msecond delay. Impulses are transmitted down through bundles of His to Purkinje fibers to the ventricles.
Heartbeat regulation	The rate of beat is modified by autonomic nervous system activity (controlled by cardiac centers in the hind-brain responding to higher brain center messages, carbon dioxide, pH, blood pressure detected by baroreceptors) and by hormones, for example adrenaline (epinephrine).
Stroke volume	The more the ventricles are filled, the more blood is ejected in diastole (Frank–Starling law).
Coronary circulation	Coronary circulation supplies the heart muscle with blood.
Related topic	Mammalian blood (C9)

Heart structure

The heart is a muscular pump whose walls are largely made up of cardiac muscle anchored in a fibrous cardioskeleton. Blood returning from body tissues enters the **right atrium** (right auricle) through two great veins, the **anterior and posterior venae cavae** (*Fig. 1*). Blood returning from the lungs enters the **left atrium** (left auricle) through **pulmonary veins**. The atria are thin-walled priming pumps which expand as they receive blood. Both contract simultaneously, pushing the flow of blood through open valves into thick-walled **ventricles**. The **right ventricle** propels deoxygenated blood into the lungs through the **pulmonary aorta** (trunk) which branches to form **right and left pulmonary arteries**. The **left ventricle** propels oxygenated blood into the **systemic aorta** whence it travels to the body tissues via the arterial system. The flow pattern is illustrated in *Fig. 2*.

The right atrium and the right ventricle are separated by the **tricuspid valve** (with three flaps) (*Fig. 3*); the left atrium and the left ventricle are separated by

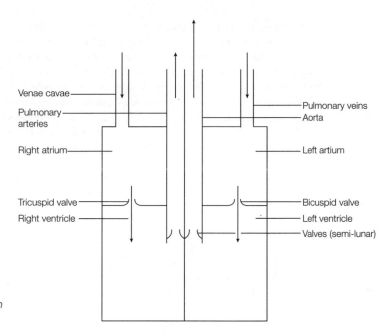

Fig. 1. Stylized
structure of the human
heart.

Venae cavae
Pulmonary arteries
Right atrium
Tricuspid valve
Right ventricle

Pulmonary veins
Aorta
Left artium
Bicuspid valve
Left ventricle
Valves (semi-lunar)

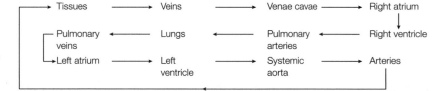

Fig. 2. The course of
blood flow in the adult
human.

Tissues → Veins → Venae cavae → Right atrium

Pulmonary veins ← Lungs ← Pulmonary arteries ← Right ventricle

Left atrium → Left ventricle → Systemic aorta → Arteries

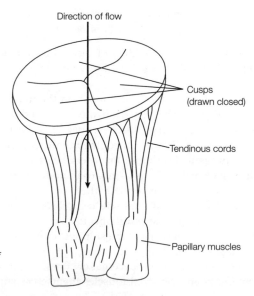

Direction of flow

Cusps
(drawn closed)

Tendinous cords

Papillary muscles

Fig. 3. The structure of
the tricuspid valves in
the human heart.

the **bicuspid** (or **mitral**) **valve** with two flaps. Each flap is thin and strong. The broad base is anchored into the fibrous cardioskeleton. Attached to the free end are tendinous cords (resembling the cords of a parachute) continuous with muscle. As blood flows into the ventricle in late diastole, the flaps lie against the ventricle walls. As the ventricle contracts in systole, the flaps come together and the atrioventricular openings are closed. The tendinous cords prevent the flaps from everting into the atria. Semi-lunar valves between the right ventricle and the pulmonary aorta (*Fig. 4*), and between the left ventricle and systemic aorta, close after ventricular contraction, preventing a back-flow of blood.

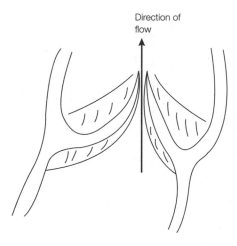

Direction of
flow

Fig. 4.   The structure of the semi-lunar valve in the human heart.

Heartbeat

In adult humans the heart beats at about 72 beats per minute. The outflow, pumping phase of ventricular contraction is **systole**, and the inflow, filling phase is **diastole**. The total volume of blood pumped by the heart is the **cardiac output**.

$$\text{Cardiac output (liters min}^{-1}) = \text{heart rate (beats min}^{-1}) \times \text{stroke volume (liters beat}^{-1})$$

e.g. 72 beats min^{-1}, ejecting 0.07 liters of blood into the aorta per beat gives a cardiac output = 72×0.07 liters = 5.0 liters min^{-1}.

Heartbeat control

Stimulation of cardiac muscle cells originates in the muscle itself: the heart will continue beating if it is removed from the body and kept in an oxygenated nutrient solution. In the embryo, the heart begins to beat early in development before it is innervated. This is called **myogenic** beating or autorhythmicity (compare the neurogenic beating of an insect heart).

The initiating **pacemaker** is the **sinoatrial node** in the right atrium: pacemaker cells can spontaneously initiate their own impulse and contract. The impulse then spreads along the muscle cells through the right and left atria, causing them to contract almost simultaneously, as the wave of excitation passes over individual muscle cells: the impulse travels very quickly across gap junctions between adjacent cardiac muscle cells.

About 100 mseconds after the pacemaker 'fires' impulses traveling through special conducting fibers, the atrial muscles themselves stimulate a second area of nodal tissue, the **atrioventricular node**. From here, impulses are carried by special muscle fibers (in the bundle of His which branches and ends as Purkinje fibers emerge) to the walls of the right and left ventricles which then contract almost simultaneously. Although the fibers in the bundle of His conduct impulses very speedily, the atrioventricular node has slow-acting fibers: thus a delay is imposed between atrial and ventricular contractions, allowing time for ventricular filling.

Heartbeat regulation

Heartbeat rate can be modified by the autonomic nervous system: parasympathetic fibers from the vagus nerve secrete acetylcholine which decreases the heart rate; sympathetic fibers secrete noradrenaline (norepinephrine) which stimulates the heart, as does adrenaline (epinephrine) from the adrenal medulla.

The activity of the nerves controlling the heart rate and the strength of the heartbeat are under the control of the cardiac centers of the medulla oblongata (hind-brain). There are separate cardioacceleratory and cardioinhibitory centers: these control sympathetic and parasympathetic innervation of the heart (and also nerve fibers controlling contraction of arterioles; see above). Thus, if there is a significant increase in blood flow in exercise due to dilation of blood vessels, the heart is simultaneously stimulated to beat faster, developing more pressure to support blood flow. The regulating center integrates reflexes controlling heart rate (and also stroke volume and force of contraction), together with blood pressure.

Raised carbon dioxide levels (and lowered pH) in exercise also stimulate the heart rate. High blood pressure (sensed by baroreceptors in arterial walls) following exercise will decrease heart rate, as will a high oxygen tension in the blood. Higher brain activity (e.g. the fight/flight/fright syndrome) leads to a release of adrenaline (epinephrine) which accelerates the heart.

Stroke volume

Stroke volume is regulated by either increasing or decreasing the force of contraction of the heart, under the control of sympathetic stimulation of the ventricles and by changes in the inflow of blood. There is a relationship between the strength of muscle contraction and the resting length of the muscle fibers. The amount of blood returned to the heart by the veins is the major factor regulating the degree of heart filling: **the more the ventricles are filled during diastole, the more blood will be ejected in systole**. This is the **Frank–Starling law of the heart**; the heart takes in and pumps out the same amount of blood during a cardiac cycle. In exercise, the body muscles massage the great veins which act as a blood reservoir – this increases inflow, stretches the heart and increases the stroke volume and force of contraction, thus aiding blood flow (and oxygen supply) to exercising tissues.

Coronary circulation

The heart muscles receive their blood from arteries branching from the aorta: this is the **coronary circulation**. The coronary arteries receive about 5% of the oxygenated blood flowing to the body. The blood drains via coronary veins to a coronary sinus which empties into the right atrium. Flow is regulated by homeostatic feedback mechanisms reflecting metabolite build-up and blood pressure.

C11 VERTEBRATE HEARTS AND ARTERIAL SYSTEMS AND THEIR EVOLUTION

Key Notes

Heart structure	Fishes have a two-chambered heart and single circulation. Amphibians and reptiles have a three-chambered heart and mixed circulation. Birds and mammals have a four-chambered heart and double circulation.
Evolution of gill arches	There is a reduction in the number of gill arches in fishes; arch 3 becomes the carotid arch, arch 4 becomes the systemic arch and arch 6 becomes the pulmonary arch.
Related topics	Phylum Chordata (A14) The mammalian heart (C10)

Heart structure

Fishes

Fishes have a two-chambered heart with one atrium and one ventricle. Deoxygenated blood is received from the body and blood is expelled through a ventral aorta to the gills. In jawed fishes, the number of gill arches reduces to four (arches 3, 4, 5 and 6) from the primitive number of six arches.

Amphibians and reptiles

Amphibians and reptiles have a three-chambered heart: the right atrium receives deoxygenated blood from the body, the left receives oxygenated blood from the lungs; both atria discharge into the single ventricle. Some separation is facilitated by spiral valves in the ventricle so that most deoxgenated blood goes to the lungs and most oxygenated blood goes to the body, although some mixing is found in more primitive forms. The ventricle is partially divided in reptiles, and is almost completely divided in Crocodilia.

Birds and mammals

The homoiothermic vertebrates possess a four-chambered heart with a parallel double circulation: the right heart pumps deoxygenated blood from the body to the lungs, the left side oxygenated blood from the lungs to the body.

Evolution of gill arches

Gill arch 3 becomes the carotid arch, gill arch 4 becomes the aorta or the systemic arch. Arch 4 is lost on the right side in mammals, and on the left side in birds; gill arch 5 is lost except in urodele amphibians; gill arch 6 becomes the pulmonary arch. The increasing separation of pulmonary and systemic circulations is notable in more advanced land vertebrates, and may be a factor permitting the evolution of homoiothermy.

C12 THE IMMUNE SYSTEM

Key Notes

Immunity	Immunity is divisible into innate (natural) immunity (nonspecific substances, phagocytic cells, integrity of body surfaces), and adaptive (acquired) immunity (adaptive, specific, xenophobic, memory).
Antigens	Antigens (e.g. pathogenic microorganisms, macromolecules) stimulate the adaptive immune response.
Evolution of immunity	Invertebrates possess features of innate immunity, whereas adaptive immunity is characteristic of vertebrates.
Two-arm response	There are two basic arms of the adaptive immune response: ● humoral (antibody-mediated) immunity, mediated by B lymphocytes, characterized by antibodies in solution; ● cell-mediated immunity, mediated by T lymphocytes, characterized by production of cytokines and effector T cells. There are complex links and interactions between the two arms.
Lymphocytes and lymphoid organs	Immunity is mediated by lymphoid cells (lymphocytes and successors) aided by accessory cells. Stem cells from the bone marrow divide in the primary lymphoid organs (thymus and bone marrow) to give virgin T and B cells respectively; these migrate to secondary lymphoid organs (e.g. lymph nodes, spleen) to await antigenic stimulation.

Related topics	Immune responses (C13)	Lymphatic system and lymph
	Hypersensitivitiy, autoimmunity and immunization (C14)	(C15)

Immunity

The **immune system** describes a set of the major defense mechanisms to protect the body against invasion by harmful agents. It is part of control and co-ordination and could be seen as an aspect of homeostasis.

Two types of immunity can be described:

(1) **Innate or natural immunity** which includes nonspecific substances present in body fluids such as tissue lysozymes and antiviral interferons (these are not usually specific and many are present before an infection), nonspecific phagocytic cells such as polymorphonuclear neutrophils ('polymorphs') and macrophages, and the integrity of the skin and the mucosae.

(2) **Adaptive or acquired immunity** which has the following features:

● **adaptivity** in that the immune response (which is previously there in potential) is called forth by an evoking **antigen**;
● **specificity** for the evoking antigen;

- **xenophobia** (= hatred of foreigners) whereby the system precisely distinguishes self and nonself;
- **memory** whereby a second exposure to an antigen calls forth a heightened and accelerated immune response.

The adaptive immune response complements the natural immune system and the various components work as a defensive whole.

Antigens

An **antigen** is any substance capable of eliciting an adaptive immune response: common antigens include large macromolecules such as proteins and nucleic acids, cells from individuals genetically different from 'self', pathogenic microorganisms such as viruses, bacteria, fungi, protoctistans and 'worms', and entities such as pollen and animal dander. The macromolecules on the surfaces of cells and pathogens determine their antigenicity.

Evolution of immunity

Invertebrates tend to possess features characteristic of the natural immune system such as phagocytic cells and nonspecific agglutinins and lysins in body fluids: there is little specificity or memory. In deuterostome animals such as starfishes and tunicates, cells morphologically similar to lymphocytes are seen and there is evidence of a rudimentary major histocompatibility complex (MHC); no true immunoglobulin antibodies are discernable.

The **vertebrates** all possess a full armory of components of the adaptive immune system such as immunoglobulin antibody molecules and an MHC. However, the responses in poikilothermic vertebrates such as agnathans (lampreys), cartilaginous fishes and urodele amphibians tend to be somewhat slow and less precise, and very temperature dependent. Teleost fishes (e.g. herrings), anuran amphibians (frogs and toads), birds and mammals seem, independently, to have evolved rapid, sophisticated immune responses (which may be the result of, or contribute to, the 'success' of these groups).

Two-arm response

Immunity can be divided into two 'arms'. **Antibody-mediated** (or **humoral**) **immunity** is associated with the appearance of **antibodies**, secreted by cells of the **B-lymphocyte** series, in the extracellular fluids such as plasma, lymph and external secretions. Antibodies are proteins, structurally described as **immunoglobulins** which, through the shapes of their binding sites, are characterized by a high physicochemical specificity for their evoking antigens.

Cell-mediated immunity is mediated by cells of the **T-lymphocyte** series with antigen-specific receptors on their surfaces. Reaction of the receptor with its antigen triggers the release of physiologically active **cytokines** or facilitates directly or, through complex signaling processes, cytotoxic activity towards the invader.

Lymphocytes and lymphoid organs

The principal mediators of adaptive immunity are **lymphoid cells** – the **lymphocytes** and their descendants of which there are about 10^{12} weighing 600 g in a healthy adult man. Lymphocytes originate as **stem cells** in the embryonic yolk-sac, the fetal liver and the adult bone marrow (*Fig. 1*). Some go to the **thymus** where they undergo a process of division, commitment (acquiring specificity for their particular antigen) and education ('learning' how they will react to antigen), partly under the influence of thymic 'hormones', and emerge as **virgin T cells**. Other stem

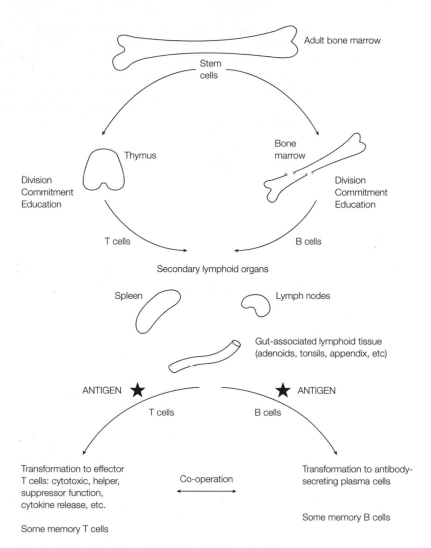

Fig. 1. Ontogeny of B and T lymphocytes in mammals.

cells undergo a similar process in the **bone marrow** itself and emerge as **virgin B cells**. [In birds, the **bursa of Fabricius**, situated in the cloaca (near the urinogenital and anal openings), acts as a B cell-maturating organ.]

The thymus and the bone marrow (or the bursa in birds) are the **primary lymphoid organs**. Thymectomy or congenital absence of a thymus results in no T cells and a grossly deficient cell-mediated immunity; other errors or ablations of the system will result in other, appropriate deficiencies in the response.

Virgin B and T cells migrate to the relevant parts of **secondary lymphoid organs** such as the **lymph nodes**, the **spleen** and **lymphoid nodules** in the lining of the gut, the respiratory and urinogenital tracts and the skin. These are strategically situated in the blood, and lymph flows to intercept antigens. Antigen, appropriately presented, stimulates division of the lymphocytes and the initiation of an immune response.

Lymph nodes are bean-shaped organs with afferent lymphatic vessels entering the convex edge and efferent lymphatics leaving the hilum (concave edge). Connective tissue extends in from the capsule forming a loose mesh allowing lymph and cells to percolate through. The outer cortex is densely packed with lymphocytes, the deep cortex (paracortex) having the T cells and the B cells being nearer the edge. The inner medulla has more loosely packed lymphocytes and macrophages.

The **spleen** has red and white cell areas or 'pulps'. Lymphocytes are in the white pulp, the T cells in cuffs around the arterioles, and B cells are outside them. Red pulp manufactures and stores erythrocytes.

C13 IMMUNE RESPONSES

Key Notes

Immune responses

Antigens are presented to B cells by antigen-presenting cells (macrophages); usually with the help of T cells. The B cells divide repeatedly then mature to form effector plasma cells (which secrete antibody) and memory B cells.

Antibodies

Antibodies are immunoglobulin (Ig) proteins with a four-chain (two heavy, two light chain) structure. Variation in amino acid sequences at the N-terminal ends of the chains gives varied shapes to the Ig molecule which match the antigen shapes, so conferring antibody–antigen specificity.

Adjunctive functions of antibodies

The C-terminal end of Ig determines biological (adjunctive) functions, e.g. precipitation and complement fixation. IgM, produced in the primary immune response, is replaced by IgG, especially in the secondary immune response. IgA is produced in secretions.

Cell-mediated immunity

Antigen presented to T cells stimulates the division of T cells to form effector T cells with varying roles including cytokine secretion. Tissue graft rejection is an example of cell-mediated immunity. Cells have MHC-encoded antigens on their surfaces, giving tissue types which are recognized by the host's immune system as nonself.

Related topics

The immune system (C12)
Hypersensitivity, autoimmunity and
  immunization (C14)

Lymphatic system and lymph
  (C15)

Immune responses

In the secondary lymphoid organs, T and B cells undergo further mitosis in response to antigen. A wide range of receptors is available on different virgin lymphocytes. The antigen 'selects' a receptor with a shape complementary to the antigen shape and stimulates that cell to divide to form a clone of daughter cells (clonal selection). Normally such division is only initiated with the aid of signaling molecules, and a complex network of signaling conversations and interactions exists between the different arms of the immune response: those interested should consult more advanced texts. Whether a response is primarily T-cell or B-cell mediated will depend on the nature of the antigen and its mode of presentation.

In **humoral immunity** most antigens are first processed by a macrophage (**antigen-presenting cell, APC**) which presents the antigen to the relevant B cell along with a signaling, 'self' MHC antigen. The B cell responds, normally with the obligatory assistance of a helper T lymphocyte; many divisions of the B cells result in a population of large **plasma cells** which secrete antibodies (together with a smaller population of small, memory B cells which will be stimulated by further exposure to the antigen and facilitate the speedier secondary immune response).

In **cell-mediated immunity**, the relevant type and specifically committed T cell is stimulated to divide; effector T cells are produced which secrete cytokines and other signaling molecules, and have cytotoxic function.

Antibodies

Antibodies (a functional term) are **immunoglobulin (Ig)** proteins (a structural term). Although all Igs are structurally related to each other, the Igs from a given animal are molecularly heterogeneous. However, antibody secreted by a single B cell or by a clone of its descendants will have the same specificity.

The basic Ig molecule has four polypeptide chains: there are two identical **heavy chains** and two identical **light chains** (*Fig. 1*). The chains are linked by chemical bonds, ionic and (usually) covalent. The protein has a higher order structure giving an overall globular shape which, when unraveled, can be described as Y-shaped. The N-terminal ends of the chains bind antigen: variation in the primary amino acid sequence of the **variable regions** of the chain (the N-terminal half of the light chains and the N-terminal quarter or fifth of the heavy chains) is reflected in the **shape of the molecule**. Such variation (encoded in a variety of groups of genes which are selected and whose products are translocated and translated together) results in an enormous variety of shapes; giving antibody diversity. The antibody shape, in its antigen-binding cleft, 'fits' the shape of the antigen: a 'lock-and-key' model. The rest of the antibody molecule is relatively **constant** in its amino acid sequence, and is responsible for the biological or **adjunctive functions** of the antibody.

The first antibody type to appear in the **primary immune response** is usually **IgM**: this is a pentamer of the basic Ig molecule, each of the five subunits having two μ heavy chains and two light chains (of which there are two types, never mixed in one Ig molecule, κ or λ). The subunits are joined by a J (joining) chain.

A few days later, the IgM is supplemented and soon supplanted by **IgG** which resembles the basic Ig molecule with two γ and two light (κ or λ) chains.

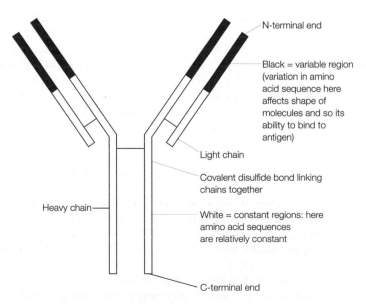

N-terminal end

Black = variable region (variation in amino acid sequence here affects shape of molecules and so its ability to bind to antigen)

Light chain

Covalent disulfide bond linking chains together

Heavy chain

White = constant regions: here amino acid sequences are relatively constant

C-terminal end

Fig. 1. Structure of an immunoglobulin protein.

Following a second injection of antigen, predominantly IgG is produced: this is the **secondary immune response**. Memory B cells are stimulated.

In secretions (e.g. gut fluids, tears, urino-genital secretions, semen, milk and colostrum), **IgA** is found. This has two basic four-chain units (each with two α and two κ *or* λ chains) joined by a J chain and stabilized by a 'secretory piece' inserted by the epithelial cells.

There are two other Ig classes (or isotypes) in humans: **IgE** (with two ϵ heavy chains and two light chains) associated with type I immediate hypersensitivity, and **IgD** (with two δ heavy chains and two light chains), an important receptor molecule on lymphocytes, particularly in early development.

In fishes, IgM is the only Ig found and is probably the most primitive in evolutionary terms. Multiple Ig classes are found in all land vertebrates: the main Ig of the secondary immune response in amphibians, reptiles and birds is **IgY** (which may be the evolutionary homolog of mammalian IgA): IgG is restricted to mammals.

Adjunctive functions of antibodies

These describe the functions of antibodies other than their binding to antigens. Among them are:

- **Precipitation:** IgG and IgM are involved in the formation of insoluble lattice complexes of antibody–antigen, ingested by macrophages.
- **Agglutination:** cross-linking of bacteria, nonself red cells, etc., by large IgM and IgA molecules. (The isohemagglutinins of the human ABO blood group system are IgMs.)
- **Complement fixation:** complement, C', is a mixture of proteins found in plasma: it can be activated by antibody binding to antigen. Through either of two pathways, C' can be 'fixed', resulting in damage to the antigen, for example bacteria are lysed. IgG and IgM are the Igs involved in this process (and IgA through the alternative C' pathway).
- **Opsonization:** IgM and IgG can coat antigens such as bacteria and make them more 'palatable' to macrophages.
- **IgG crosses the placenta** in many mammalian species, including humans.
- **IgA is found in external secretions**, for example gut fluids. IgA in milk and colostrum can confer passive immunity to the mammalian neonate in many species.
- **IgE** is responsible for features of **immediate (type I) hypersensitivity** (see Topic C14).

Cell-mediated immunity

An example of cell-mediated immunity is the **rejection of a graft** of genetically different, nonself tissue (e.g. skin or kidney). In very much simplified terms, T lymphocytes 'see' nonself antigens on graft (donor) cells. All cells have transplantation or tissue-type antigens on their surfaces: these are highly specific to individuals and are encoded in the **MHC** genes. T cells are activated by the nonself antigens and, with the assistance of accessory cells, transform, secrete **cytokines** and facilitate in various ways the destruction of the graft. The generation of memory T cells during this process results in the speedier rejection by the 'sensitized' host of a second ('second-set') graft from the same donor although a graft from another donor will be rejected by the host with the same kinetics as the first graft from the first donor. (There will also be antibodies made to graft antigens.)

Grafts will not be rejected if:

- the graft is not vascularized;
- the host and donor have the same genotype and so have the same MHC genes (e.g. identical twins or members of inbred strains of mice);
- the host has been made tolerant (specifically and actively *non*reactive to the donor's cells (e.g. by giving the host very small amounts of antigen before grafting);
- the graft has been 'matched' for donor and host MHC products (tissue types);
- the host is immunologically deficient due to immaturity, senescence, disease or immunosuppression (e.g. by surgery, drugs or irradiation).

Graft types

- Autografts: from one part of an individual to another.
- Isografts: from one individual to a genetically identical individual.
- Allografts: between individuals of the same species.
- Xenografts: between individuals of different species.

C14 HYPERSENSITIVITY, AUTOIMMUNITY AND IMMUNIZATION

Key Notes

Hypersensitivities
Hypersensitivities are immune 'overreactions' which are damaging to self, for example IgE-mediated anaphylactic hypersensitivity in hay-fever and T cell-mediated delayed hypersensitivity to tuberculin.

Autoimmunity
Autoimmunity is a reaction to self-antigens, for example rheumatoid arthritis, Hashimoto's autoimmune thyroiditis, antisperm antibody following vasectomy.

Immunization
Passive immunity is conferred by maternal IgG crossing the human placenta to the fetus, or through maternal IgA in milk. Active immunity is conferred by giving killed antigen, attenuated antigen or antigen extract/parts artificially to induce a protective immune response.

Related topics
The immune system (C12)   Lymphatic system and lymph (C15)
Immune responses (C13)   Aging (D8)

Hypersensitivities A vigorous immune reaction can result in tissue damage to 'self'; such reactions are termed **hypersensitivities** and could be seen as the 'cost' of an efficient immune system. Four types of hypersensitivity reaction are usually recognized:

- Type I   Anaphylactic, atopic 'immediate' hypersensitivity
- Type II   Cytotoxic hypersensitivity
- Type III   Complex-mediated hypersensitivity
- Type IV   Cell-mediated 'delayed' hypersensitivity

Types I and IV are briefly described. Many hypersensitivities are related to occupations, for example farmer's lung is a type III hypersensitivity introduced by fungal spores in moldy hay.

Type I (atopy and anaphylaxis)
IgE fastens by its C-terminal end to mast cells (or basophils). Allergens (such as pollen in hay-fever) act as antigens and bind to the variable, N-terminal region of the IgE. This alters the antibody shape, the mast cell membranes are activated and vasoactive amines such as histamine are released from the cells. Within a few minutes ('immediately') these can cause an **anaphylactic reaction**, including smooth muscle contraction, inflammation, etc. This reaction is the basis of hay-fever, some types of asthma, certain food allergies, etc.

IgE-mediated reactions may have a natural 'use' in dislodging worm parasites from the gut.

Type IV [cell-mediated (delayed) hypersensitivity]
This hypersensitivity relies not on antibodies but on T lymphocytes. One type of this reaction involves patients who have become sensitized to tuberculin, found in *Mycobacterium tuberculosis*, the causative agent of tuberculosis. An injection of tuberculin into the skin results in tuberculin ingestion by macrophages and their presentation of it to T cells which infiltrate into the area after about 48 hours ('delayed'), causing a hard, red swelling (induration and erythema).

Other type IV hypersensitivity reactions are a primary cause of cavitation in the lungs in tuberculosis, and are the basis of a number of allergies such as those to stainless-steel jewellery or hair dyes.

Autoimmunity

In autoimmunity the body has, through a variety of complex ways, become immune to 'self'. Sometimes the immune system is exposed to antigens from which it was previously shielded, for example many vasectomized men make antibodies to their own sperm. In other cases, antigens combined with drugs may elicit responses which act against the uncombined antigen itself. Further autoimmune reactions may be the result of breakdowns in the mechanisms whereby the immune system censors components which react against self.

Many autoimmune diseases are found more commonly in old age, for example rheumatoid arthritis and a number of autoimmune thyroiditis conditions (e.g. Hashimoto's disease). As with hypersensitivities, autoimmunity may be a price to be paid for having an efficient immune system.

Immunization

Immunization involves the induction of immunity to protect an individual. **Passive immunization** involves giving pre-formed antibodies to a person (or animal): these antibodies are gradually broken down by the normal processes of protein catabolism, so this form of immunity is temporary. Passive immunization can be natural or artificial.

Natural passive immunization includes the passage of maternal IgG across the mammalian placenta. In regions with a high incidence of tetanus, immunization of pregnant women with tetanus toxoid results in them making antitetanus IgG which is transmitted across the placenta to the fetus, and can protect the newborn from tetanus. In humans, IgA is transmitted to the baby's gut via colostrum and milk.

Artificial passive immunization is effected when immunodeficient patients are given doses of antibodies from a donor; travelers to the tropics may be given 'pooled γ-globulins' (antibodies) from donors who live in the visited areas; hopefully, the cocktail of donated antibodies may protect them from endemic diseases. There is a danger that recipients will mount immune reactions to the antibodies themselves (which they will 'see' as 'nonself' proteins), resulting in immune complexes depositing themselves on the glomerular membranes of the kidneys (this is an example of type III complex-mediated hypersensitivity). Genetic engineering techniques are being pursued to develop high-affinity antibodies against rubella (German measles virus), etc.

Active immunization (vaccination) works on the principle of stimulating the individual's own immune system to make antibodies, effector cells and memory T and/or B cells in response to the infectious agent, usually delivered in a dead or less virulent form. The aim is to provide effective immunity by establishing adequate levels of antibody and a population of memory lymphocytes which can expand rapidly on renewed contact with antigen. For poliomyelitis protection, high levels of blood antibody are desirable; in mycobacterial diseases such as tuberculosis, a macrophage-activating cell-mediated immunity is sought; cytotoxic T cells are most effective with influenza infection. The site of the response may be important – for cholera the response must be in the gut cavity to present adherence to and colonization of the gut wall. Antigens need to be cheap, safe (not injuring the recipient) and stable. Some types of antigen used are:

- Killed organisms (e.g. cholera, typhoid, paratyphoid A and B).
- Live attenuated organisms: these replicate and give a more sustained dose of antigen. Attentuation can be achieved by using strains of pathogen virulent in one species but avirulent in another (e.g. Jenner's cowpox which protects against smallpox), or by modifying growth conditions (e.g. Calmette and Guérin who modified *Mycobacterium tuberculosis* by adding bile to the growth medium to produce the BCG vaccine). Attenuated vaccines for polio (Sabin), rubella and measles have gained wide acceptance.
- Individual antigens [e.g. tetanus and diphtherial exotoxins (detoxified with formaldehyde to yield a toxoid)]; pneumococcal and meningococcal polysaccharide vaccines are usually coupled to a strongly antigenic carrier protein. Genetic engineering allows genes to be inserted into a vector so that the gene for an immunogenic protein is expressed on the vector surface.

C15 LYMPHATIC SYSTEM AND LYMPH

Key Notes

Lymph	Tissue fluid between cells is lymph, a plasma-like fluid containing leukocytes. The lymph is drained by the lymphatics.
Lymphatics	The lymphatic system contains blind-ending capillaries which in mammals join progressively to form lymphatic thoracic ducts which empty into the subclavian veins.
Lymph nodes	Lymph nodes, important secondary lymphoid organs, are situated in the lymphatic system.
Lymphatic circulation	In mammals lymphatic circulation is by body muscle contraction. Lymph hearts may be present in lower vertebrates.
Related topic	The immune system (C12)

Lymph

In mammals, most tissue fluid is withdrawn back into the blood by osmosis ('pulled' by the colloidal osmotic pressure of the plasma) at the venous end of capillaries. Some excess fluid in the interstitial (intercellular) spaces is not returned to the blood, nor is fluid which may escape from the blood when vessels are damaged (e.g. by a blow), and a pool of fluid (an edema) collects. Such fluids lost by the blood are collected by the **lymphatic system** which routes them, as lymph, back to the bloodstream.

Lymphatic system

The lymphatic system consists of a network of progressively larger vessels (**lymphatics**): the smallest vessels are blind-ending capillaries, the largest are similar to veins. They do not form part of a continuous circuit. Tissue fluid seeps into the lymph capillaries from whence it travels to large **lymphatic thoracic ducts** which drain into the subclavian veins near the collar bones. The lymph consists of a plasma-like fluid (in which the protein concentrations may differ from those found in blood plasma) and cells, particularly leukocytes: erythrocytes are not normally present. Specialized lymphatics (**lacteals**) are found in the villi of the gut: these are involved in fat transport to the liver.

Lymph nodes

Lymph nodes are distributed throughout the lymphatic system: they are situated strategically to intercept antigens and are important secondary lymphoid organs.

Lymphatic circulation

In mammals, lymph is propelled by body muscle contraction and perhaps by some peristaltic contraction of the lymphatics. Valves prevent back-flow and low pressure in the thorax on inspiration aids flow.

In lower vertebrates, lymphatic capillaries are often open-ended. Lymph circulation can be assisted by **pulsatile lymph hearts**.

C16 OSMOREGULATION

Key Notes

Osmoregulation	Osmoregulation is the control of electrolyte and organic solute concentrations in the body fluids, and the maintenance of water balance and fluid volume.
Body fluid compartments	The body fluid compartments are intracellular and extracellular (interstitial fluid, body cavity fluids, blood, lymph).
Extracellular fluid	Body fluids contain electrolytes and organic solutes. The total solute concentration determines the osmolarity of fluid. Cell membranes act like semi-permeable membranes. Important regulated ions include sodium, potassium, calcium, chloride and protons. Protons are important in the regulation of the acid–base balance: excess protons are eliminated by excretory organs.
Osmoregulation and the environment	Osmoregulation relates to the salt concentration of the environment, the degree of desiccation (especially in air), the water needs of physiological processes, and to the need to eliminate wastes.
Aquatic invertebrates	Most marine invertebrates are osmoconformers, but many selectively regulate ions. Some marine crustaceans are osmoregulators. All freshwater invertebrates are hyperosmotic to their medium.
Aquatic vertebrates	Aquatic vertebrates can be isosmotic (elasmobranchs) or hyposmotic (teleosts) to the sea; all are hyperosmotic to fresh water. Marine elasmobranchs and crab-eating frogs retain urea in their blood to maintain isosmolarity. Marine teleosts lose water in the sea, but compensate by drinking and void excess salts actively through their gills. Freshwater teleosts gain water from their environment, lose it through copious urine, and actively take up salt through gills. Freshwater amphibians are similar, taking up salts through the skin.
Terrestrial animals	Moist-skinned terrestrial animals burrow, live in damp habitats, retreat to shells or use slime. Terrestrial arthropods have waterproof exoskeletons to minimize water loss, but some is lost through respiratory organs. Water is resorbed from urine and feces. Reptiles, birds and mammals have waterproof skins. Marine reptiles and birds drink sea water but excrete a concentrated salt solution through their nasal glands; marine mammals excrete hyperosmotic urine using long loops of Henle in kidneys, as do desert mammals which can use metabolic water.
Related topic	Nitrogenous excretion (C17)

Osmoregulation

Osmoregulation is concerned with the regulation of body fluid solute concentration in the animal and the maintenance of water balance.

Body fluid compartments

Most animals comprise mainly water: a 60 kg adult man contains about 40 liters of water, 25 liters within cells and the rest in extracellular fluids. The extracellular body fluids include:

- Interstitial water between cells
- Body cavity fluids (e.g. in coelom)
- Blood
- Lymph

The extracellular body fluids need to remain in osmotic balance with the fluid in cells from which they are separated by the cell membrane. This may include adjustment of electrolyte concentrations or organic solute concentrations (which may differ on each side of the membrane: intracellular fluid has a higher potassium concentration than extracellular fluid, whereas the reverse applies for sodium). It is important that such differences are maintained, particularly in excitable cells such as neurons. Extracellular fluids are routes for the transport of gas, food, waste, etc. to and from cells. The body fluids need to be in osmotic balance with external fluids in the case of aquatic animals, or some means of control must be supplied: this may be critical in cells closest to the external environment (as in the respiratory, urino-genital and alimentary tracts).

Extracellular fluid

Total solute concentration determines the **osmotic pressure** of the solution. Cell membranes act as semi-permeable membranes, and increases in the total solute concentration of extracellular fluids may result in damaging withdrawal of water from cells.

Sodium (Na^+) is the main cation and **chloride (Cl^-)** is the main anion, and together they are two of the main contributors to the fluids' osmotic pressure; changes in Na^+ and Cl^- content promote water retention or excretion in order to maintain the osmotic pressure of the extracellular fluids. They are, therefore, key components in maintenance of fluid and thus of body volume.

Potassium (K^+) is found principally within cells; fluctuation in extracellular K^+ affects the electrical properties of nerve and muscle cell membranes.

Calcium (Ca^{2+}) is critical in determining the excitability of nerve and muscle cells; about 50% of the plasma calcium is bound to plasma proteins, the rest is ionized in solution and is under close control.

Proton (H^+) concentration (measurable as pH) is important in acid–base regulation in the body; the main sources of protons are from amino acid metabolism and from carbon dioxide.

$$CO_2 + H_2O \rightleftharpoons H_2CO_3 \rightleftharpoons H^+ + HCO_3^-$$

Excess base (e.g. HCO_3^-) can combine with protons to remove them from solution. The high reactivity of protons renders it imperative that they be regulated in all body fluid compartments. Buffers such as hemoglobin and phosphate in cells, and bicarbonate/carbonic acid outside cells fulfill this role. Blood plasma pH is about 7.35–7.45; intracellular pH varies between 6 and 7 (due to acid metabolites in the cells). Protons are also excreted in the urine: thus the proton concentration in extracellular fluids is kept constant. (In some lung disorders where carbon dioxide is retained, there is a build-up of plasma proton levels as more HCO_3^- is generated; protons are removed by the kidneys and retention of bicarbonate ions by the kidneys increases the buffering capacity of the blood.)

Voiding of water can be also used secondarily to void waste products, or can be a primary method of voiding those products. Thus osmoregulatory and excretory organs are often combined.

Regulation of the composition and volume of body fluids may include consideration of the following factors.

- The salt content of the environment in which an aquatic animal lives, and the necessity to maintain an internal salt balance. In fresh water, water will tend to flow into the animal by osmosis and expulsion of excess water will result in loss of salts; in salt water, water may flow out of the animal and compensatory drinking of salty water may result in salt overload.
- In terrestrial species there may be a tendency to desiccate in dry air.
- The water requirements of physiological processes (e.g. digestion, sweating).
- The need to eliminate nitrogenous wastes (which will require water which may be in short supply).

Osmoregulation and the environment

Animals live in the sea, in brackish water, in fresh water and on land: all pose osmoregulatory problems. Seventy one percent of the earth's surface is covered with water, 1% is fresh water (0.01% volume of sea water).

The sea is 3.5% salt, the major ions being Cl^-, Na^+; additionally Mg^{2+}, SO_4^{2-} and Ca^{2+} are present in substantial amounts. (There are geographic variations: the Mediterranean has 4% salts, coasts and estuaries may be less saline, although relative proportions of ions stay approximately the same.) Some inland waters may be very saline: the Great Salt Lake, Utah, USA is saturated with NaCl: there are no fishes in this lake but the brine-shrimp *Artemia* sp. survives. The Dead Sea in the Jordan Rift Valley is saturated with several ions, and $CaSO_4$ crystallizes out: only microorganisms survive. Brackish waters near estuaries are interesting as a transition site between freshwater and saltwater life forms. Fresh waters are very varied with respect to their solute content: droplets of ocean spray evaporate, and salt particles can be carried far inland and deposited with rain; leaching of salts from rocks occurs (e.g. Ca^{2+} from limestone).

Animals can be classed as:

- **Euryhaline**: tolerate wide salinity variation.
- **Stenohaline**: intolerant of wide salinity variation.

Aquatic invertebrates

Most **marine invertebrates** are **isosmotic**: as **osmoconformers** they change their body fluid osmolarity to match the concentration of the medium; the concentration of solutes may differ from the medium, even if the osmolarity is the same (because of organic solvents); therefore, there may be **ionic regulation**. The osmoconforming jellyfish *Aurelia* sp. regulates to give low SO_4^{2-} levels in its body fluids, giving it positive buoyancy. Mg^{2+} (which anesthetizes neuromuscular transmission) is down-regulated in fast-moving crab species.

Osmoregulators maintain the osmolarity of body fluids despite changes in the medium: many marine crabs in brackish water maintain a high body fluid osmolarity. Osmoregulators survive better in **brackish water** than do osmoconformers, although some are intermediate. The mitten crab *Eriocheir* sp. can survive fresh water but returns to the sea to breed; *Carcinus* sp. can only survive if salts are at 33% of the sea water concentration. The **brine-shrimp**

Artemia sp. is normally hyposmotic to its medium: salts from food are removed from the gut by active uptake and then eliminated by active transport across the gill epithelia.

All **freshwater invertebrates** are hyperosmotic to the medium and osmoregulate, although there are differences between concentrations of body fluids, that of a crayfish, *Astacus* sp., being five times that of a freshwater clam, *Anodonta* sp. Such freshwater animals face problems of osmotic inflow of water, especially through respiratory organs, and also through the gut. Excess water can be voided as dilute urine, but no animal is known which can void pure water so there will be a loss of salts. These can be compensated for by salts in food or by direct uptake from the medium, for example through the gills in crayfishes by active transport processes.

Aquatic vertebrates

Aquatic vertebrates can be:

- **Isosmotic** (or slightly hyperosmotic) with respect to sea water (e.g. hagfishes, elasmobranchs, lungfishes, *Rana cancrivora*).
- **Hyposmotic** with respect to sea water (e.g. lampreys, teleosts).

In **fresh water** all vertebrates are **hyperosmotic**.

Hagfishes (one of the two types of living jawless vertebrates) are marine and stenohaline; that is their salt concentration is equal to that of sea water. All other vertebrates keep their salts at a concentration lower than that of the sea. **Elasmobranchs** (sharks and rays) are almost all marine. They are in osmotic equilibrium with the sea by adding urea and trimethylamine oxide (TMAO) to their body fluids; the shark kidney resorbs urea. Their plasma urea concentration is 100 times that of mammals. Excess salts are excreted into the gut by a rectal gland. (There are a few freshwater elasmobranchs which all have low plasma urea concentrations, suggesting that this system is labile in evolution.) **Holocephali** (ratfishes) and the **coelacanth** also have high plasma urea and TMAO levels, using a similar osmoregulatory strategy to elasmobranchs. **Dipnoi** (lungfishes) are freshwater forms which normally void ammonia via the gills; estivating dipnoans retain urea (to a concentration of 4%) in the blood, although this is associated more with estivation rather than being osmoregulatory.

The only sea-going **amphibian**, *Rana cancrivora*, the crab-eating frog from brackish mangrove swamps in the Mekong delta, also retains urea in its blood, giving it a blood slightly hyperosmotic to the sea: thus water flows in slowly through the skin, giving water for urine and obviating the need to drink salty water; the tadpole osmoregulates like a teleost, but the frog does need to seek fresh water to lay its eggs and for the period of metamorphosis. Freshwater amphibians maintain their plasma concentrations at about one-third that of sea water. Osmotic inflow of water is compensated for by voiding dilute urine: salt loss is restored by active transport of ions through the skin.

Teleosts, both marine and freshwater, are hyposmotic to the sea (approximately one-third of the sea water concentration).

- **In fresh water** they gain water through their gills and lose water by voiding dilute urine. Loss of salts in urine is compensated for by active uptake of ions through the chloride cells of the gills (*Fig. 1*).
- **In the sea** they lose water via their gills and compensate by drinking. Excess salt is excreted by active transport through the chloride cells in gills (*Fig. 1*).

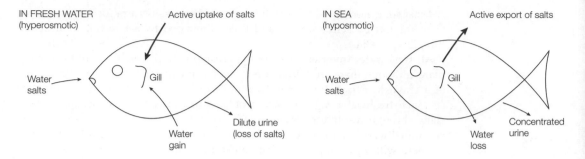

IN FRESH WATER (hyperosmotic)   Active uptake of salts

Water salts

Gill

Water gain

Dilute urine (loss of salts)

IN SEA (hyposmotic)   Active export of salts

Water salts

Gill

Water loss

Concentrated urine

Fig. 1.   Osmoregulation in teleosts with body fluids at approximately one-third the osmotic concentration of the sea.

Skin in marine teleosts is very permeable to ions while skin in freshwater teleosts is much less permeable. The killifish adapts to both environments. The permeability of skin to ions drops in a few minutes when they are placed in fresh water, but recovery of permeability when they are placed in sea water takes some hours. **Anadromous** fishes (e.g. salmon) can live in both sea and fresh water. The osmotic flow of water can reverse, and active ion exchange at the gills can change direction; two separate populations of chloride cells are probably involved.

Terrestrial animals

Moist-skinned animals tend to desiccate in air: the dryer the air the greater the saturation deficit (i.e. the difference in vapor pressure of water in air over a free water surface and vapor pressure in air) and the more likely the animal is to dry out. The process is exacerbated by air movement (wind). Thus, moist-skinned animals tend to live in moist microhabitats (earthworms, newts, frogs), and/or have shells in which to retreat (snails), and/or have protective mucus (slugs). Water loss is compensated for by drinking.

The Australian desert toad *Chiroleptes platycephalus* estivates in burrows several feet deep: very dilute urine is stored in the bladder. After rain, it surfaces and lays eggs which rapidly hatch, develop and metamorphose in transient puddles. Lungfishes are similar: in drought they make a coccoon with a breathing duct to the surface; they then estivate – protein metabolism results in high plasma urea concentrations. The desert snail *Sphincterochila* sp. withdraws into its shell and secretes a calcareous epiphragm over the entrance, allowing it to survive mid-summer sunshine; during such a period of estivation water loss can be less than 500 μg per day (total water = 1.5 g).

Animals with waterproof skins do not face some of the difficulties encountered by moist-skinned forms; however, respiratory organs are often a source of water loss. Most terrestrial **crustaceans** need to return to water to lay eggs: *Geocarcinus* sp. can take up water from moist sand. Land woodlice need to dip their gill-books in water to keep them moist.

Insects and arachnids are very successful on land: the tracheae are lined with chitin and only the terminal tracheoles are water-permeable; 'cyclic' respiration whereby spiracles only open for brief periods while carbon dioxide is 'blown off' from the tissue fluids also conserves water. In some species the cuticle can take up water; damage to the waterproof wax coat of the cuticle can lead to a fatal dehydration. Many insects are very effective at resorbing water from urine (see Topic C17) and from feces; storage of excretory products (e.g. crystals of uric acid in the fat body) also conserves water. Metabolic water (produced during the oxidation of foodstuffs) can be utilized, and a few insects seem to be able to absorb water from air (e.g. the flea *Xenopsylla* sp. can absorb water from air with a relative humidity of at least 50%).

Reptiles possess thick, scaly, waterproof skins. Many desert species can tolerate fluctuations in blood electrolyte concentrations; for example *Trachysaurus* sp. can tolerate a blood sodium concentration of 150% normal. Marine reptiles such as the Galapagos marine iguana, crocodiles or sea-snakes drink sea water. Urine is hypotonic to the sea, so excess salts are excreted from nasal salt glands. Those feeding on teleosts ingest food which is hypotonic to sea water (see above). All **birds** have paired nasal salt glands too: in marine birds the salt glands are enlarged and secrete a hyperosmotic NaCl solution intermittently.

Marine mammals drink sea water, but their possession of kidneys with very long loops of Henle (see Topic C17) means that they can produce urine hypertonic to sea water. Thus there is a net gain of water to the seal or whale when it drinks sea water:

	Sea water drunk	Urine produced	Water balance
Human	1000 ml 535 mosm l⁻	1350 ml 400 mosm l⁻	−350 ml
Whale	1000 ml 535 mosm l⁻	650 ml 820 mosm l⁻	+350 ml

Such hypertonic urine and long kidney loops are also used as a water conservation ploy by **desert mammals** (e.g. jerboa); many of these mammals are not normally seen to drink water and rely on metabolic water gained from oxidation of foodstuffs.

C17 NITROGENOUS EXCRETION

Key Notes

Nitrogenous excretion	Nitrogenous excretion is removal of the nitrogenous waste products of metabolism.
Nitrogenous wastes	The principal nitrogenous waste is highly toxic ammonia. To conserve water it is detoxified to either urea or uric acid.
Invertebrate excretory organs	Invertebrate excretory organs are saccular, receiving filtrate directly from the blood, or tubular, opening directly to the coelom. Crustaceans have a green gland: fluid is filtered from the hemolymph into a saccule. Insects have Malpighian tubules: waste is secreted into a tubule, and salts and water are withdrawn from the rectum.
Vertebrate kidney	Most vertebrates have ultrafiltration saccular kidneys with selective resorption and secretion superimposed. In mammalian kidneys the ultrafiltrate has the same small solute concentrations as blood; glucose, salts and water are resorbed in the proximal tubule. The countercurrent multiplier in the loop, and the collecting tubule allow the production of hypertonic urine.
Regulation of urine	Antidiuretic hormone (ADH) permits increased resorption of water in the collecting tubule so that more concentrated urine is produced. Aldosterone facilitates sodium recovery in the kidney. Low blood pressure results in renin/angiotensin release leading to vasoconstriction which helps to maintain the filtration rate.
Evolution of the vertebrate kidney	The vertebrate kidney has evolved from an ancestral segmental holonephros. Anterior parts are progressively lost or taken over by the male reproductive system.
Related topics	Homeostasis (C1) Osmoregulation (C16)

Nitrogenous excretion	Nitrogenous excretion is the removal of nitrogenous waste substances of metabolism.
Nitrogenous waste substances	During **protein catabolism (breakdown)** amino acids are deaminated to form a keto-acid and **ammonia**. Ammonia is very soluble in water and, as it goes into solution, some combines with protons in the water to form NH_4^+. This is very toxic, but is quickly flushed from the body if copious water is available (500 ml of water will flush away 1 g of nitrogen). Most fishes excrete ammonia as do amphibian tadpoles. Terrestrial animals need to conserve water so they detoxify ammonia to **urea** in the liver. This saves about 90% of the water needed

to void 1 g of nitrogen. **Urea** is the main nitrogenous excretory product of mammals. Further detoxication of ammonia yields the purine **uric acid**, the principle nitrogenous waste product of land snails, insects, reptiles and birds. This only requires less than 10 ml water to eliminate 1 g of nitrogen, but uric acid is metabolically expensive to produce.

Invertebrate excretory organs

Protoctistans use a **contractile vacuole**. In *Paramecium* sp. there is one at each end of the ciliate, each composed of a radiating ring of tubules which deliver water to a central vesicle; on reaching a certain size, the vesicle discharges (by fibrils contracting and then by hydrostatic pressure) through a pore in the outer pellicle.

In **Metazoa** a few animals void ammonia by **simple diffusion** to the exterior (e.g. cnidarians). Most animals employ excretory organs which can be **tubular or saccular** (*Fig. 1*). In **tubular organs** the excretory tubule opens into the coelom. Fluid is filtered from the blood into the coelomic fluid whence it passes into the tubule: here selective resorption of salts, sugars, amino acids and water may occur, and secretion of wastes by the tubule wall takes place, helped by the blood vessels adjacent to the tubule. The final solution produced is the **urine**. **Saccular organs** have a closed tubule which does not open into the coelom: the tubule receives filtrate directly from the blood. The filtrate composition is modified by absorption and secretion to yield urine.

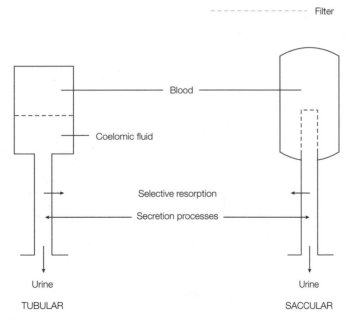

Fig. 1. *Tubular and saccular invertebrate excretory organs. Saccular organs include protonephridia of flatworms and tubular organs include the metanephridia of molluscs.*

In many animals the osmoregulatory/excretory tubules are termed **nephridia,** which open to the exterior by a nephridiopore. Flatworms and some marine annelids have **protonephridia**. The inner, blind end of this saccular organ has a terminal cell with 'flickering' cilia, a **flame cell**. The wall of the flame cell has microvilli where it joins the next cell of the tubule, and the resulting slits function as a filter. The beating of the cilia creates a negative pressure and interstitial fluid is drawn into the tubule through these slits. Molluscs and most annelids possess **metanephridia**. The inner end is not blind but has a **nephrostome** funnel opening into the coelom. Earthworms possess one pair of such tubular organs per segment.

Green (antennal) glands are found in **crustaceans** such as crabs. The organ has a saccule in the head, bathed in hemolymph. The saccule has interdigitating podocyte cells allowing the filtration of fluid from the hemolymph (the cells are also found in protonephridia and in vertebrate kidneys). From the saccule, fluid passes to a labyrinthine tubule where selective resorption occurs; the resulting urine is stored temporarily in a bladder in some species before being expelled via a pore at the base of the antenna. (The saccule is a remnant of the coelom: thus the tubule may be homologous to a metanephridium.)

Insects and spiders possess **Malpighian tubules**, blind-ending tubules bathed in hemolymph (compare Crustacea) (*Fig. 2*). There are from two to several hundred tubules which open into the intestine. Filtration does not take place into the tubules: uric acid (as soluble potassium urate), potassium and sodium are secreted from the hemolymph into the tubules and water follows. The contents then pass down to the rectum. Ions are actively pumped out of the fluid and again water follows; uric acid crystallizes out. Thus hypertonic urine (particularly in species from arid habitats) is produced and water is conserved.

(a)

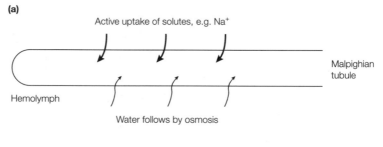

(b)

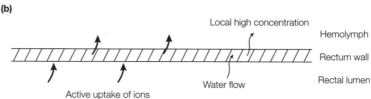

Fig. 2.   Insect excretion and water conservation in (a) Malpighian tubules leading to (b) the rectum.

Vertebrate kidney   Most vertebrate kidneys work on the principle of **ultrafiltration** of plasma (leaving cells and macromolecules behind) from the high-pressure, closed blood system, with elements of selective solute and water **resorption** and **secretion** superimposed.

The filtration/resorption system can process large quantities of fluid, but often 99% of the fluid is resorbed: a secretory kidney would seem to imply less energy expenditure. Some **teleost** fishes have **exclusively secretory kidneys**; however, this does not allow them to eliminate novel toxins effectively in a changed environment, Ultrafiltration is nonselective with respect to small molecules. All vertebrate kidneys can produce urine hypotonic or isotonic with respect to the blood; only birds and mammals can make hypertonic urine.

Each kidney is made up of several million **nephron** units comprising glomeruli and kidney tubules (*Fig. 3*). Ultrafiltration (filtration powered by the hydrostatic pressure of arterial blood) occurs in the **Malpighian corpuscle**: a knot of capillaries (the **glomerulus**) is embedded in the hollow cup-shaped end of the kidney tubule, the **Bowman's capsule**. The ultrafiltrate passes down a **proximal convoluted tubule**, round the hairpin bend of the **loop of Henle**, through the **distal convoluted tubule** to the **collecting duct**. The collecting ducts from all the tubules drain into the **renal pelvis** from which urine passes out via the the **ureter**, often being stored temporarily in the bladder.

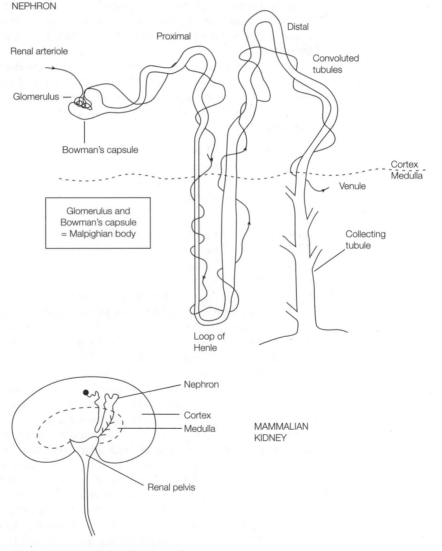

Fig. 3.   *The mammalian kidney.*

Blood pressure forces fluid out from the thin walls of the glomerular capillaries: the hydrostatic pressure of the arterial blood must exceed the colloidal osmotic pressure of the plasma proteins (if the renal artery is clamped, ultrafiltration ceases). In amphibians (which have Malpighian bodies near the kidney surface), the ion, urea and glucose concentrations in the Bowman's capsule are the same as in plasma. **Inulin** is not metabolized, resorbed or secreted, but is filtered: if urine inulin is at 100 times its plasma concentration, then 99% of water in the ultrafiltrate must have been resorbed. In a normal man, the filtration rate is more or less constant at approximately 130 ml per minute, 780 ml per hour, about twice the plasma volume; thus, resorption of water must be very efficient.

Glucose is actively resorbed in the proximal convoluted tubule: normal plasma glucose levels are about 100 mg per 100 ml; if this is exceeded the amount of glucose filtered and resorbed can rise to 230 mg per ml. This is the tubular maximum, T_M. Above this level, plasma glucose spills over into the final urine, resulting in glucosurea, as in diabetes mellitus. Salts are also resorbed, and water follows (about 85% of filtrate water is resorbed here in humans).

Tubular secretion eliminates phenolics and their detoxification products, penicillin and other xenobiotics. Phenol red excretion has been used to determine renal blood flow: it is removed completely by the kidney. Phenol red can be injected into the bloodstream, and the amount appearing in urine after a given time noted.

Adult man: renal plasma flow   = 0.7 liters per minute
            hematocrit   = 45%
            therefore blood flow = 1.25 liters per minute
            (approximately one-quarter of the resting cardiac output)

Hypertonic urine in birds and mammals

Birds can make urine with up to twice the plasma concentration of salts; mammals up to 25 times (desert mammals are most successful at this). The process is based on Na^+ and Cl^- resorption, using the loop of Henle (*Fig. 4*): the concentrating ability is related to the length of the loop. Most mammals have nephrons with long and short loops; desert species (e.g. jerboa) have only long loops, aquatic species (e.g. beaver) have mainly short loops.

Ultrafiltrate fluid enters the loop with a Na^+ concentration, $[Na^+]$, equal to the plasma concentration (150 mosm l^{-1}). Chloride is actively transported out of the ascending arm of the loop and there is positive movement outwards of sodium ions; this increases the $[Na^+]$ in the surrounding tissue to, say, 200 mosm l^{-1}. The descending arm is permeable to sodium and to water, and sodium diffuses in until the $[Na^+]$ in the descending arm is equal to that in the surrounding tissue (200 mosm l^{-1}). The ascending arm must be water-impermeable to prevent water passively following the sodium ions into this arm.

Fluid flowing out of the loop into the distal tubule is sodium depleted. Therefore the $[Na^+]$ is less than that in the plasma. As the fluid which leaves the loop has less sodium than that which entered, sodium must accumulate and concentrate in the loop in higher and higher concentrations: the longer the loop, the higher the concentration. This is a **countercurrent multiplier system**. The fluid entering the collecting tubule which parallels the loop is very dilute: the wall is permeable and water flows out by osmosis into the high $[Na^+]$ of

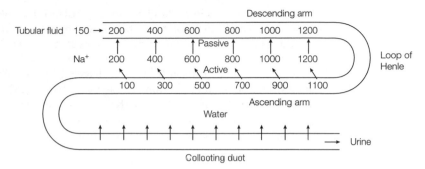

Fig. 4. *The generation of hypertonic urine using the loop of Henle. The numbers show the sodium ion concentration (arbitrary units); chloride ion concentrations would be similar.*

the surrounding tissues. This leads to concentrated urine. Urea also plays a role in the concentration gradients.

Birds are less effective at producing hypertonic urine: they excrete uric acid, and too concentrated a uric acid solution would clog the tubules. There is some evidence that they resorb water in the cloaca. Marine birds eliminate excess salt using nasal glands (see Topic C16).

Regulation of urine

Human urine flow can vary from 10 to 1000 ml per hour. About 85% of water is withdrawn in the proximal tubules as a consequence of salt and water resorption. The distal convoluted and collecting tubules are responsible for the concentration of urine: the process is partly controlled by **ADH (vasopressin)**. ADH is secreted by the hypothalamus, passing via the posterior pituitary to the bloodstream. ADH acts by increasing the water-permeability of the collecting ducts.

● When the body is overloaded with water, ADH secretion ceases and the collecting tubules become impermeable to water, leading to copious, dilute urine (diuresis).

● When the body is dehydrated, ADH is secreted making the collecting tubules more permeable to water, leading to small volumes of concentrated urine.

The osmotic concentration of blood is sensed by **osmoreceptors** in the hypothalamus: **shrinkage of cells** stimulates ADH secretion (and also a **thirst urge**). (ADH has a similar effect on frog skin, making the skin more permeable to water and stimulating osmotic inflow of water when desiccated.)

If the body is overloaded with water but low on salts ADH secretion ceases (low plasma osmolarity) and maximum recovery of sodium occurs in the distal tubule. This is stimulated by **aldosterone** secretion by the adrenal cortex: it accelerates the sodium pumps in the distal tubule. If aldosterone is accompanied by ADH secretion, water follows sodium isosmotically out of the tubule and into the blood. If ADH is absent (but aldosterone is present), the tubule is waterproof: sodium is extracted from the tubule leaving water behind. Thus aldosterone facilitates sodium removal and conservation.

Aldosterone itself is stimulated by a **decrease in blood pressure** in afferent arterioles of the kidney (e.g. following hemorrhage) and decreased [Na⁺] in the early part of the distal tubule. When the blood pressure falls, **juxtaglomerular apparatus (JGA)** cells release **renin**. Renin acts on **angiotensinogen** (from the liver) to convert it to **angiotensin I** and this is later converted to an octopeptide **angiotensin II**. Angiotensin is a powerful vasoconstrictor, so blood pressure rises and glomerular filtration rates are restored It also stimulates the synthesis and release of aldosterone (see above), and so sodium retention. Thus the kidney acts as a **homeostat** to maintain its own glomerular filtration rate. (Decreased sodium in distal tubules is detected by macula densa cells in the tubule wall, also leading to renin release by JGA cells, aldosterone and sodium retention. If [Na⁺] falls relative to [K⁺], aldosterone is released, and sodium pumps in the distal tubules exchange sodium in the filtrate for potassium in the plasma.) This scheme is summarized in *Fig. 5.*

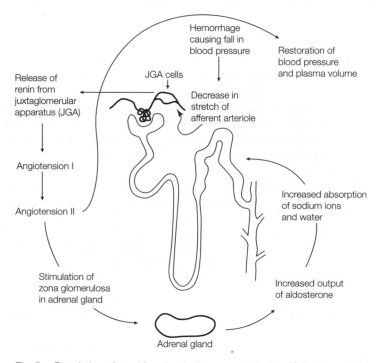

Fig. 5.   Regulation of renal function in the mammal by the JGA cells and aldosterone.

Evolution of the vertebrate kidney

Ancestral vertebrates probably possessed a segmental kidney, with one pair of nephrons for each body segment between the anterior and posterior ends of the coelom. They drained into an **archinephric duct**. Such a kidney is a **holonephros** (*Fig. 6*).

In adult fishes and amphibians, the anterior tubules are lost, the middle tubules are associated with draining the testis in the male, and the functional kidney is the hind-most part of the holonephros: this is known as an **opisthonephros**. The archinephric duct functions in most amphibians and fishes as a ureter and a sperm duct (teleost fishes have a separate sperm duct).

Reptiles, birds and mammals (amniotes) have a secondarily nonsegmental **metanephros** with its own ureter. Most of the old opisthonephros is now (in males) the epididymis associated with the testis, and the archinephric duct becomes the vas deferens.

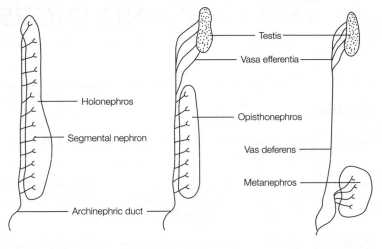

Fig. 6.   Vertebrate kidney evolution.

During amniote embyrogenesis, there is a process which partially parallels this evolutionary sequence: there are transitory pronephros and mesonephros in the front and middle portions of the embryonic nephrogenic mesoderm. The most posterior part develops to form a metanephric kidney.

C18 THE GUT AND DIGESTION

Key Notes

Feeding patterns	Feeding patterns include macrophagy, microphagy, fluid feeding and osmotrophy.
Digestive system regions	Guts can be generalized into five regions: reception; conduction and storage; trituration and early digestion; further digestion and storage; formation, compaction and elimination of feces.
Human gut structure	The human alimentary tract has mucosa, submucosa, muscularis externa and serosa layers.
Mouth	The teeth and tongue masticate food. Saliva lubricates and has an amylase.
Esophagus	The esophagus transports food to the stomach.
Stomach	Gastric juices in the stomach have acid to kill bacteria and break down fibers. Pepsin initiates digestion of proteins.
Small intestine	The alkaline environment of the small bowel allows continuation of digestion by pancreatic enzymes and enzymes on the surfaces of epithelial cells. This is aided by the emulsifying salts in bile. The large surface area (due to villi and microvilli) aids absorption of digestion products.
Large intestine	Water resorption occurs in the large bowel.
Co-ordination of digestion	Digestion is under the control of the autonomic nervous system and hormones: gastrin in stomach, secretin and cholecystokinin–pancreozymin in the small bowel.
The mammalian liver	Hepatocytes produce bile which contains fat-emulsifying salts, cholesterol and hemoglobin breakdown products. Bile is temporarily stored in the gall-bladder. The liver has many other functions associated with homeostatic control and metabolism.
The mammalian pancreas	The pancreas secretes digestive hormones which flow into the intestine via the pancreatic duct and the common bile duct. The islets of Langerhans are an endocrine organ producing hormones associated with blood glucose control.
Ruminant digestion	Ruminant mammals have symbiotic microorganisms in esophageal sacculations to break down (principally) cellulose in the diet. Lagomorphs use symbionts in the cecum.

Mammal tooth patterns	Mammalian dentition patterns reflect diet, for example grinding teeth in herbivores, especially grazers, and tearing, cutting teeth in carnivores.
Related topic	Human blood glucose control (C19)

Feeding patterns

Macrophagous feeders (eat large food particles) include carnivores which eat meat (includes fish-eating piscivores, and insect-eating insectivores, etc.), herbivores which eat vegetation and omnivores which eat a mixed diet. They often have jaws and teeth.

Microphagous feeders include suspension feeders (e.g. annelid fanworms or urochordate sea-squirts) which use netting devices to filter and concentrate small particles suspended in water (filter-feeding) and other devices such as mucus strings and parachutes.

Fluid feeders include forms such as aphids tapping fluid in plant phloem.

Osmotrophs include forms such as tapeworms and other parasites absorbing nutrients through the body surface.

Digestive system regions

Digestive systems show great variation in the Animal Kingdom. Gut structures do not always correlate with feeding habits (although large stomachs may reflect large, infrequent meals). There is a close correlation between gut structures and the nature of the ingested food: carnivores tend to have short absorptive intestines.

A five-part gut is recognizable.

(1) **Reception** by the mouth and pharynx. Mechanical trituration can occur here (e.g. use of tongue and teeth in mammals); enzymic digestion may commence; toxic factors may paralyse prey. Fluids lubricate the food.

(2) **Conduction and storage** in the esophagus which may be distended to form a crop (birds) or fermentation chambers (ruminant mammals).

(3) **Trituration and early digestion** in the stomach and the first part of the intestine. Mechanisms for internal trituration may be found in animals which swallow large food masses [e.g. mammalian stomach muscles; muscles and gravel in bird gizzards; rotation of a crystalline style in the bivalve stomach (*Fig. 1*)]. In crustaceans, a chitin-lined gizzard precedes the

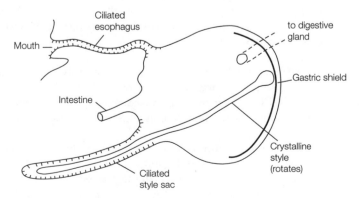

Fig. 1. Section through the stomach region of a bivalve mollusc, e.g. Donax *sp.*

intestine: food is ground in a gastric mill with dorsal and lateral teeth moved by muscles. Enzymic digestion is commonly found in this region. Diverticula are common (e.g. teleost fishes): cilia may propel food into diverticula in invertebrates (e.g. molluscs).

(4) **Absorption and further digestion** occurs in the vertebrate small intestine: most of the enzymes come from the pancreas. The liver secretes bile. Some absorption occurs in the duodenum but mostly in the ileum. In most invertebrates, digestion has occurred in stomach diverticula and the intestine is absorptive. The area is increased by typhlosole (longitudinal fold, e.g. earthworm), spiral valve (elasmobranchs), folding and villi (e.g. mammals).

Area of the human intestine:

Simple tube	$3.3 \times 10^3 \text{ cm}^2$
With folds	10^4 cm^2
With villi	10^5 cm^2
With microvilli	$2 \times 10^6 \text{ cm}^2$

(5) **Compaction and formation of feces** where much water is absorbed in terrestrial species (particularly important in insect osmoregulation).

Human gut structure

The rather unspecialized human gut reflects our omnivorous diet. The inner surface of the gut is continuous with the exterior of the body so the gut **lumen** is technically outside of the body. The gut is lined by a **mucosa** (epithelial tissue, underlying basement membrane, connective tissue with thin smooth muscle), a **submucosa** (outside the mucosa with connective tissue, blood vessels, nerves), **muscularis externa** (with inner circular muscles and outer longitudinal muscles) and an outer **serosa** of connective tissue. Co-ordinated muscle contractions produce ring-like contractions to churn up food and waves of peristalsis to propel food along the gut.

Mouth

Food enters the gut via the **mouth**; mastication by the teeth and the tongue takes place. Saliva, which contains bicarbonate and an α-amylase, lubricates the bolus of food. Approximately 1500 cm^3 saliva per day are secreted by the salivary glands; secretion is under parasympathetic nervous control. Human teeth are relatively unspecialized; each jaw quadrant has two biting incisors, one tearing canine, two premolars and three molars for grinding. Teeth are coated with enamel over dentine: in the middle is a vascularized, innervated pulp. Teeth are embedded in jaw bones by ligaments and cement. Absorption of some drugs occurs in the mouth (e.g. morphine).

Esophagus

The triturated, lubricated food bolus passes down the **esophagus** by peristalsis; the esophagus is lined (like the mouth) with stratified squamous epithelium. Peristalsis is movement of a muscular tube, such as the gut, by co-ordinated contractions of smooth muscle in a definite (usually anterio-posterior) direction.

Stomach

Food is triturated further in the **stomach** by bands of smooth muscle – longitudinal and circular muscles antagonize each other. This allows mixing with gastric juices. Food first collects in the relaxed body of the stomach; pronounced peristaltic waves from the top (fundus) to the base (antrum) occur – this generates **chyme** which is squirted through the pyloric sphincter in small quanta. The sphincter is normally open except after a squirt into the duodenum (i.e. it acts as a valve: a large meal can take more than 4 hours to enter the duodenum).

The stomach mucosa is very thick, with many gastric pits. Mucus-secreting cells cover the stomach surface and line the pits. Gastric juice is very acid with hydrochloric acid (HCl, pH = 1.5–2.5), which is secreted by gastric glands in the lower parts of the pits: acid kills bacteria and living cells in the food, loosens fibrous components in food and facilitates conversion of pepsinogen to pepsin. Protein digestion is initiated in the stomach: gastric juice contains pepsinogen (an enzyme precursor or zymogen) which is converted to pepsin by gastric HCl. (Zymogens prevent digestion of the stomach by self-enzymes. Mucus also protects cells from activated enzymes; the mucus forms a protective coat absorbed onto a **glycocalyx** on the surfaces of microvilli of lining cells.) Rennin precipitates soluble proteins in milk. Water, salts, some vitamins and some drugs (e.g. ethanol) are absorbed in the stomach.

Small intestine

The small bowel (intestine) comprises, in sequence, the duodenum, jejunum and ileum: it is the site of enzymatic digestion using enzymes from the pancreatic juice and in or on the surfaces of intestinal epithelial cells. Pepsin is inactivated by mildly alkaline conditions in the small bowel. Protein digestion continues using pancreatic trypsin and chymotrypsin. Pancreatic juice also has an amylase and a lipase. Bile salts from the liver (draining into the duodenum from bile ducts after temporary storage in the gall-bladder; see below) emulsify fats: bile also acts as an excretion medium for hemoglobin degradation products and cholesterol.

Much of the final breakdown of proteins, fats and sugar polymers occurs within cells lining the small bowel: as intestinal epithelial cells mature and pass up to the tips of the villi, the enzyme contents increase. It is now thought that digestion occurs in the cells or on their membranes – there does not seem to be a secretion of large quantities of enzymes into the small bowel lumen (in which those enzymes which are found derive from the pancreas).

Digestion products are absorbed in the small bowel whose area is increased by villi and microvilli (see above). Some fats, hydolyzed to fatty acids and glycerol, are resynthesized to new fats, and are packaged into chylomicrons and absorbed by lacteals (of the lymphatic system) in the villi: chylomicrons deliver fats to the adipose cells or liver or are broken down in the bloodstream. Cholesterol is made in the liver and is packaged into low-density lipoprotein (LDL) complexes which are excreted in bile. Stored cholesterol can be repackaged for delivery to cells for membrane or steroid hormone synthesis.

Large intestine

Absorption of water, sodium and other minerals occurs in the **large bowel (intestine).** The first part is the colon where, in humans, 7 liters of water are absorbed per day. The colon harbors bacterial flora which can further digest food and synthesize absorbable amino acids and vitamins (e.g. vitamin K). The cecum, almost absent in humans, is a blind pouch with the vermiform appendix at its end; it is a secondary lymphoid organ. Residues (dead cells, bacteria, cholesterol, bile pigments, undigested food, especially cellulose fibers) form the feces, stored in the colon. Feces are expelled periodically by passing them down the rectum and out through the anus.

Co-ordination of digestion

Enzymes act consecutively along the the gut at specific sites: this requires release of small quantities of food and a precise sequence of enzyme release. Food movement is largely under **autonomic nervous control**. **Saliva secretion** is nervous alone: it is initiated by food in the mouth, or by the smell, sight or anticipation of food.

Gastric juice secretion is initiated by nervous stimulation due to food in the mouth and/or the stomach. Distension of the stomach antrum results in the hormone **gastrin** being secreted by stomach-lining cells into the bloodstream, stimulating further gastric juice secretion (an isolated, denervated stomach pouch still secretes gastric juice when food is in the antrum). Gastrin secretion is inhibited by high acid levels in the stomach (negative feedback). As stomach contents are released into the duodenum, the duodenum secretes a gastrin-inhibitory polypeptide (GIP) which further inhibits gastric juice secretion.

Acid stomach contents in the duodenum stimulate release of **secretin**: this enhances the flow of bicarbonate-rich (alkaline) pancreatic juice. Protein fragments stimulate **cholecystekinin–pancreozymin (CCK–PZ)** which leads to the release of pancreatic enzymes and stimulates gall-bladder contraction. **Motility hormones** affect movement of the villi by smooth muscle contraction. *Fig. 2* illustrates the interacting hormones in this system.

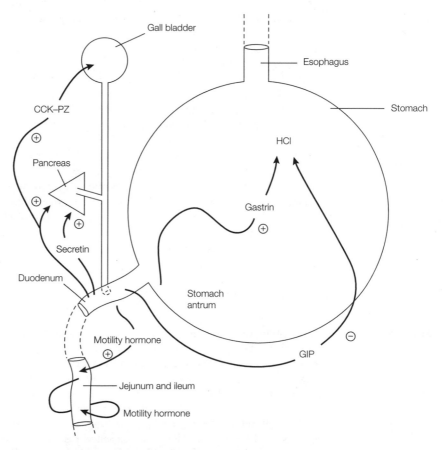

Fig. 2.   Hormonal control of digestion in a mammal.

The mammalian liver

The liver develops as an outgrowth of the gut; most of its cells are **hepatocytes** arranged in cylindrical lobules intimately associated with venules, arterioles and bile canaliculi. Food-enriched blood is brought from the gut to the liver by the hepatic portal vein, and oxygenated blood is transported to the liver by the hepatic artery. The liver is drained by the hepatic vein.

Liver cells secrete **bile** which passes down bile canaliculi into a hepatic duct and then up a cystic duct to the gall-bladder. Relaxation of a sphincter at the neck of the gall-bladder and contraction of the bladder following the release of food into the duodenum allows bile to flow down the common bile-duct into the small intestine. Bile acts as an excretion medium for cholesterol and hemoglobin degradation pigments, but its main role is to carry bile salts which emulsify fats in the bowel.

Other functions of the liver include glycogen, amino acid and fat storage and metabolism, detoxification of ammonia to urea, cholesterol synthesis, fetal erythropoiesis (red cell manufacture), breakdown of excess hemoglobin, red blood cell storage, vitamin storage, synthesis of many plasma proteins and heat production.

The mammalian pancreas

Like the liver, the pancreas develops as an outgrowth of the small intestine, its cells producing large quantities of digestive enzymes which travel to the gut via the pancreatic duct, which joins the common bile duct just before its junction with the bowel. The islets of Langerhans are important endocrine glands producing hormones (e.g. insulin, glucagon) associated with blood glucose control (see Topic C19).

Ruminant digestion

Ruminants, for example cows and sheep, have a gut divided into sections similar to those of humans, except that the esophagus and stomach are greatly modified. They use symbiotic bacteria, yeasts and protoctistans to break the β,1–4 links in cellulose. There are four chambers: the rumen, reticulum and omasum are sacculations of the esophagus; the abomasum is the true stomach.

On being swallowed, food goes to the reticulum (tripe) where it is made into cud balls. The fermenting mass is regurgitated to the mouth for further trituration (chewing the cud). On second swallowing, food passes to the rumen. The rumen is rich in anaerobic symbiotic microorganisms (especially bacteria and ciliates) which ferment cellulose to fatty acids, carbon dioxide and methane, and starches to sugars. Fatty acids are absorbed by the glandular epitherlium of the rumen, the gases are eructed ('burped'!). The fermented mass then passes to the omasum (psalterium): here it is further triturated, strained and squeezed by strong muscular contractions. It then passes to the abomasum where digestive juices start work: bacteria are also digested in the abomasum and are rich sources of nitrogen and vitamins of the B complex.

Lagomorphs, for example rabbits, have a large cecum and appendix off the colon. This contains bacteria which digest cellulose and produce B vitamins. Lagomorphs re-ingest fecal pellets, giving food a second passage through the gut and permitting the microorganisms and the products of their cellulose digestion to be absorbed.

Mammal tooth patterns

Basic ('primitive') pattern
The basic pattern is **three incisors, one canine, four premolars and three molars** on each side, in each of the upper and lower jaws. The dental formula is I 3/3 C 1/1 PM 4/4 M 3/3.

Variations

(1) *Insectivora (e.g. shrew)*: unspecialized teeth, similar to the 'primitive' pattern.
(2) *Rodents (e.g. squirrel)*: one incisor, no canines or anterior premolars. Diastema between the incisor and back premolars allows cheeks to be drawn in, closing off the front of the mouth. There is enamel only on the front edge of the incisors: dentine wears away leaving a chisel edge. Incisors grow, as do premolars and molars. A longer length of articulation with the skull allows backward and forward movements. There is a nonrigid join between the mandibles: thus two incisors can move relative to each other, for example when opening nuts. All these variations facilitate gnawing.
(3) *Lagomorphs (e.g. rabbit)*: similar to rodents, but two incisors.
(4) *Perissodactyls (e.g. horse)*: three incisors but no canines; the first premolar is vestigial. There is diastema and the molars and premolars are square. The high-crowned teeth, which grow throughout life, have rough, grinding surfaces.
(5) *Artiodactyls (e.g. cow)*: cows have lost all their upper incisors. They crop grass using lower incisors and horny upper gums. A muscular and protrusible tongue sweeps grass into the mouth. They have stout, elongated molars with rough grinding surfaces (enamel, dentine and cementum wear away at different rates, so hardened cementum and enamel form ridges).
(6) *Proboscideans (e.g. elephant)*: the nose and upper lip form a prehensile trunk. One upper incisor grows throughout life to form the tusk (for defense). There are no canines or adult premolars; three molars per jaw, used one after another, develop in a series from front to back. Molars consist of many plate-like 'cones' joined by cementum producing a hard working surface with transverse ridges.
(7) *Primates (e.g. monkey, ape or human)*: omnivores with low-crowned, unspecialized teeth. They possess two incisors, one canine, two or three premolars and three molars.
(8) *'Edentate' groups (e.g. anteaters, pangolins)*: they have a tendency to lose teeth in evolution. They possess a long tongue with adhesive saliva.
(9) *Carnivores (e.g. cat, seal)*: land forms (fissipedes) have piercing, cutting incisors, large tearing, slashing canines, cutting and shearing molars and premolars (last upper premolar and first lower molar form shearing carnassials). Articulation allows an up-and-down bite. Marine forms (pinnipedes) have peg teeth which hold prey until it is swallowed (the walrus has tusks to dig up molluscs and its flattened molars can crush).
(10) *Cetaceans (e.g. whales, dolphins)*: whale-bone whales have baleen plates to filter-feed krill; toothed whales have many peg-like teeth to hold prey until it is swallowed (compare seals).

C19 HUMAN BLOOD GLUCOSE CONTROL

Key Notes

Blood glucose norm	The blood glucose norm is approximately 80–110 mg per 100 ml.
Pancreatic hormones	The islets of Langerhans in the pancreas secrete insulin, glucagon and somatostatin.
Insulin	Insulin is released from β cells of the pancreas in response to high glucose loads. Receptors occur on the liver, on fat and on muscle. This prevents inactivation of glycogen synthetase *a*.
Glucagon	Glucagon is released from α cells of the pancreas. Release is stimulated by low insulin, fasting and stress. Glucagon promotes glycogenolysis and gluconeogenesis in the liver.
Somatostatin	Somatostatin is released from δ cells of the pancreas; it inhibits insulin and glucagon in a paracrine tight feedback loop. This is local control.
Other control systems	Control is exercized by the autonomic nervous sytem, adrenaline (epinephrine) and by glucocorticoids.
Diabetes mellitus	Diabetes mellitus occurs when there is a failure of glucose control. The juvenile-onset condition is insulin dependent, and there is a lack of β cells. The late-onset condition is noninsulin dependent, and there is an autoimmune loss of insulin receptors.

Related topics	Homeostasis (C1)	Hormones (C22)
	The gut and digestion (C18)	

Blood glucose norm

Blood glucose control is part of the energy management of the body and is a key homeostatic process. The normal blood glucose load is 4.5–6 mmol l^{-1} (80–110 mg 100 ml^{-1}). Following a high carbohydrate meal, the glucose load stays about the same; an overnight fast will lower it by only about 20%. Departures from the norm are severe in their effects:

● A decrease to less than 3 mmol l^{-1} leads to hypoglycemia, causing nausea, cold sweat, loss of concentration; coma follows resulting in inadequate fuel to the central nervous system (CNS).

● An increase to more than 9 mmol l^{-1} leads to hyperglycemia, producing acidosis, glycosurea and coma.

Food interrelationships

- Glycogen → glucose (glycogenolysis)
- Lactate, alanine, glycerol → glucose (gluconeogenesis)

Fatty acids tend to be β-oxidized and fed into the tricarboxylic acid (TCA) cycle.
   A build-up of acetyl-coenzyme A (acetyl-CoA) from β-oxidation of fatty acids tends to lead to ketone-body formation; for example β-hydroxybutyrate. Excess ketones (e.g. in diabetes mellitus) lead to coma.

Pancreatic hormones

Digestion products are absorbed in the gut and carried to the liver. The **pancreas** near the liver is important for control. About 98% of pancreatic cells produce digestive enzymes, the remaining 2% of the cells, in the **islets of Langerhans**, produce controlling hormones. α cells make glucagon; β cells make insulin; δ cells make somatostatin.

Insulin

Insulin lowers the blood glucose load. Of the islet cells, 60% are β cells: these release insulin, a 51 amino acid polypeptide. Insulin is stored in islets cells in granules bound to zinc. It is released by exocytosis in response to blood with a high glucose load flowing through the pancreas. Autonomic innervation also leads to insulin release, as in anticipation of food or when food reaches the mouth and/or stomach. Calcium channels are activated by glucose (there is no insulin release in a Ca^{2+}-free medium). Calcium binds to calmodulin which promotes fusion of the insulin-containing vesicles in the cell membrane and the secretion of insulin.
   Insulin receptors are found on liver, fat and muscle cells. When insulin interacts with its receptors glycogenesis is stimulated, resulting in lowered blood glucose levels. Insulin prevents the conversion of active glycogen synthetase *a* to inactive glycogen synthetase *b*. (In obesity, levels of glucose and insulin are often both very high: tissues seem insensitive to insulin, perhaps due to reduced numbers of receptors. Fasting can increase the number of insulin receptors.)

Glucagon

Glucagon antagonizes insulin by increasing the blood glucose load. α cells make up 25% of islet cells: these release glucagon, a 29 amino acid polypeptide. Glucagon is stimulated less by low glucose, more by stress, fasting or low insulin levels. In periods of low-glucose/high-glucose demand (e.g. exercise, cold), insulin levels are low and glucagon pulls glucose out of store. It acts on liver cells, where glycogen phophorylase is activated and glycogen synthetase deactivated. In the absence of insulin, fat and protein breakdown occurs, liberating fatty acids, glycerol and amino acids. Glucagon stimulates amino acid uptake by the liver and generally promotes gluconeogenesis. (Prolonged fasting leads to desensitization of liver cells to glucagon, preventing continous breakdown of fat and structural protein.)

Somatostatin

Somatostatin is made by the hypothalamus and by islet δ cells: it inhibits growth hormone release. In the pancreas, somatostatin inhibits both insulin and glucagon, that is it has a paracrine (local intercellular messenger control) role. Glucagon release from α cells stimulates insulin release through activating cyclic adenosine monophosphate (cAMP) in β cells, and promotes secretion of somatostatin by δ cells. Therefore, a tight feedback loop is used. Somatostatin probably prevents Ca^{2+} movement in α and β cells; thus calcium-dependent release of glucagon and insulin is inhibited.

Other control   *Autonomic nervous control*
systems   There are numerous nerve endings in the islets; noradrenaline (norepinephrine)
                           stimulates glucagon secretion, acetylcholine stimulates insulin. Insulin and
                           glucose receptors are found in the hypothalamus, a key area for temperature
                           regulation and feeding/satiety control, so the hypothalamus commands the use
                           or storage of glucose. Higher brain input is also important: expectation of a meal
                           elicits insulin secretion (feedforward); stress results in lowered insulin but raised
                           glucagon levels. Nervous control overrides 'lower' local pancreatic controls.
                              Adrenaline (epinephrine) inhibits insulin, activates glycogen phosphoryl-
                           ase and promotes glycogen breakdown; glucocorticoids act on the liver to de-
                           activate glycogen phosphorylase and to promote synthesis and preservation of
                           glycogen.

Diabetes mellitus   Approximately 2% of humans have **glucose control failures**: a common failure
                           is a high blood glucose load (e.g. >180 mg 100 cm^{-3}) and glycosurea (glucose
                           in urine).
                              There are various possible **causes** (i.e. a multiple etiology):

                           - Absence of β cells
                           - Insensitivity of β cells to glucose
                           - Overactivity of α cells
                           - Overactivity of δ cells
                           - Lack of insulin receptors on target cells
                           - Inability to store glycogen
                           - Supersensitive glucagon receptors

                           Two types of diabetes:

                           - **Juvenile-onset**, insulin-dependent diabetes: onset is usually at less than 20
                             years old. People with this form of the disease respond to insulin treatment.
                             They usually lack β cells. There is no glucose clearance, glycogenolysis is
                             extensive and there is no check on glucagon. Fats are hydrolyzed, leading
                             to free fatty acids which are converted to ketones (pear-drop breath).
                             Injections of insulin are indicated for treatment. Loss of β cells may be due
                             to viral infections.
                           - **Late-onset**, noninsulin-dependent diabetes. This is arguably an inherited
                             condition. People with this form of the disease respond to glucose meals by
                             insulin secretion, show no ketosis and have lowered levels of insulin recep-
                             tors. Treatment is by controlling carbohydrate intake: insulin falls and β cells
                             are receptive again. Autoantibodies to insulin receptors are possibly impli-
                             cated.

C20 TEMPERATURE RELATIONSHIPS

Key Notes

Temperature relationship definitions	Animals may be defined as cold-blooded and warm-blooded; poikilotherms and homoiotherms; or ectotherms and endotherms, but these definitions have their limitations.
Effects of temperature change	Q_{10} measures the increase in metabolic rate over $10^{\circ}C$: the value is usually between 2.0 and 3.0 for most animals.
Temperature extremes	Animals may be stenothermic (tolerate narrow temperature ranges) or eurythermic (tolerate broad temperature ranges). Heat death may be caused by an imbalance in Q_{10} values for interrelated reactions, or thermodenaturation of enzymes. Cold death may be due to cessation of metabolic reactions. Cold tolerance is due to supercooling of body fluids, tolerance of ice crystals or the presence of antifreezes.
Related topics	Homeostasis (C1) Thermoregulation (C21)

Temperature relationships definitions

Animals are limited to fairly narrow temperature ranges (approximately $-2^{\circ}C$ to $+50^{\circ}C$ for internal temperatures which will sustain active metabolism). Most animals (including all aquatic invertebrates) have the same temperatures as their surroundings. Birds and mammals keep their body temperatures approximately constant, and many other animals maintain a substantial difference between their own body (or parts of their body) temperature and that of the environment.

Definitions

- **Warm-blooded** (e.g. mammal) – but a hibernating mammal may be at $4^{\circ}C$.
- **Cold-blooded** (e.g. lizard) – but a lizard basking may be at more than $40^{\circ}C$.
- **Poikilothermic** (= changeable temperature) – but deep-sea fish may be at a very constant temperature throughout life.
- **Homoiothermic** (= constant temperature) – but this may change in a hibernating hedgehog: the term **heterothermic** may be preferable for hibernating homoiotherms.
- **Ectothermic** – relies on external sources of heat (e.g. lizard, but neonatal mice may be temporary ectotherms).
- **Endothermic** – relies on internally generated, metabolic heat (e.g. bird, but shark and tuna muscles and moth flight muscles may conserve metabolic heat).

Thus these definitions obviously have their limitations.

Effects of temperature change

Within limits, raised temperatures accelerate most physiological processes: oxygen consumption increases as the metabolic rate of the animal increases. Within physiological limits, a rise of 10ºC leads to an increase in oxygen consumption (and therefore of metabolic rate) by a factor of 2–3. The **increase in rate over 10ºC = Q_{10}**. Therefore, if $Q_{10} = 2$, the rate doubles over 10ºC. Thus, if the rate $= x$ at 0ºC, it will be $2x$ at 10ºC, it will be $4x$ at 20ºC, it will be $8x$ at 30ºC and so on; that is, there is an exponential increase.

In practice, Q_{10} values are usually between 2 and 3; for example for a metabolic rate in a Colorado beetle between 7 and 30ºC, $Q_{10} = 2.17$. (The Q_{10} value falls at higher temperatures, particularly as enzyme denaturation starts to occur.) By contrast, the Q_{10} values for physicochemical reactions are generally less than 1.2.

Temperature extremes

Some animals can tolerate only a narrow range of temperatures (**stenothermic**), others a wide range (**eurythermic**). Tolerances can change with time (with less tolerance when immature). There is a distinction between temperatures at which an animal can survive and that at which it can carry out its entire life cycle. The lethal temperature is often difficult to define and exposure time can be important, as for humans in a sauna bath.

High temperature tolerance
There is no known animal which can complete its life cycle at more than 50ºC, but resting stages can be very tolerant: the larva of the fly *Polypedium* sp. can dehydrate and survive 102ºC for 1 minute. *Triops* sp. eggs (a Sudanese crustacean) can estivate in mud during drought at 80ºC and will survive short periods at even higher temperatures in the laboratory. Humidity is important: cockroaches will survive for 24 hours at 37ºC in damp air but will die at the same temperature in dry air.

Many animals die when exposed to temperatures above 50ºC. Marine forms would be exposed to temperatures greater than 50ºC only if they were in the intertidal zone. For snails, heat resistance reflects their position in the intertidal zone. The freshwater desert pupfish from Devil's Hole, California lives in water at a constant 34ºC – its **upper lethal limit** is 43ºC; the antarctic ice-fish *Trematomus* sp., living in water at -1.9 ± 0.1ºC, has an upper lethal limit of 6ºC. **Heat death** is probably due to different Q_{10} values for interdependent metabolic reactions: changes in temperatures unbalance the ordered sequence of reactions and the proportions and availability of (intermediate) products. Extreme high temperatures lead to enzyme denaturation.

Low temperature tolerance
Some animals will tolerate freezing, but the **lower lethal limits** for others is higher: a guppy kept at 23ºC dies at 10ºC – the respiratory centers of the hindbrain become anesthetized, leading to anoxia.

Cold and freezing tolerance may be due to:

● **Supercooling** (cooling below 0ºC at atmospheric pressure without freezing): water can be supercooled, dependent on the temperature, time and the absence of nuclei for ice crystal formation. Solutes may enhance supercooling. It may be important for a lizard exposed to –6ºC when caught out on a cold night.
● **Freezing tolerance**: a few animals can tolerate ice crystals (and the resulting high concentrations of salts). *Chironomus* larvae will survive repeated

freezing and thawing to –25ºC. It may be important for intertidal organisms in winter polar seas. Ice crystals are outside of the cells which are distorted. Freezing tolerance varies with season.

● **Antifreezes**: glycerol is found in some insect hemolymphs. It may protect cells against ice crystal damage, lower freezing points and promote super-cooling. In the wasp *Bracon cephi*, glycerol concentrations rise to 30% in the autumn, giving a freezing point of –17.5ºC; supercooling permits survival below –47ºC. In teleost fishes, supercooling is noted and there is evidence for antifreezes. In *Trematomus* sp. the antifreeze is a glycoprotein.

Tolerance limits will vary according to acclimatization (artifically induced acclimatization = acclimation), season, age, geographical distribution, etc. Arctic temperatures are lethal to tropical species as are tropical temperatures to polar species. Even closely related species (e.g. *Rana* spp.) have tolerances related to their geographical distributions (*Fig. 1*). The teleost bullhead has an upper lethal limit of 28ºC in winter and 36ºC in summer. If bullheads are acclimated to a changed temperature, the shift in temperature tolerance is rapid when the temperature of acclimation is moved upwards, but slower when it is moved downwards.

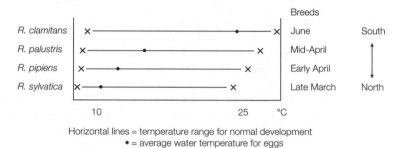

Horizontal lines = temperature range for normal development
● = average water temperature for eggs
✕ = lethal limits

Fig. 1.   *Temperature limits and tolerance with regard to geographical distribution in* Rana *sp.*

C21 THERMOREGULATION

Key Notes

Thermoconformers and thermoregulators	Animals may be thermoconformers or thermoregulators (using intrinsic, physiological or extrinsic, behavioral regulatory strategies).
Body temperature in mammals and birds	Mammals and birds are homoiotherms with varying shell (peripheral) temperatures but relatively constant core temperatures; core temperatures may fluctuate diurnally. Thermoregulation is under sophisticated autonomic nervous control.
Heat transfer	Heat may be lost or gained by conduction, convection, radiation or evaporation; heat may also be gained through metabolism.
Heat balance	For homoiothermy, heat loss, Q = heat gain, H. $Q = C(T_B - T_A)$, where C is a constant measuring insulation, etc., T_B is body temperature and T_A is ambient (environmental) temperature.
Keeping warm	In order to keep warm, the metabolic rate or insulation can increase. The area:volume ratio is important. In the thermoneutral zone, the metabolic rate stays the same with lowering ambient temperature (insulation, etc. increases); below the lower critical temperature, the metabolic rate increases – the lower critical temperature is lower in well-insulated polar species. Terrestrial mammals/birds insulate with fur/down feathers, whereas marine homoiotherms use subcutaneous blubber.
Keeping cool	For keeping cool, the area:volume ratio is critical. The main heat loss is through evaporation – sweating, panting, salivating over limbs, wallowing in mud.
Thermoregulation in poikilotherms	Thermoregulation is seen in poikilotherms: behavioral thermoregulation occurs in basking lizards; localized endothermy is seen in tuna muscles and moth flight muscles.
Torpor and hibernation	Small mammals and birds may lower their T_B in diurnal torpor or winter hibernation; the metabolic rate is radically reduced. This process is under the close control of a hypothalamic 'thermostat'.

Related topics	Homeostasis (C1)	Temperature relationships (C20)

Thermo-conformers and thermoregulators	Most animals are **thermoconformers** (whose body temperature conforms to that of the surrounding environment) but some animals are **thermoregulators**, either **behaviorally** or **physiologically**.

Body temperature in mammals and birds

Heat produced by an animal must be transported to the surface before it can be transferred to the environment. Thus the surface is at a lower temperature than the core or there would be no transfer: the animal is not at a uniform temperature. In humans, organs of the thorax and the abdomen, together with the brain, generate 72% of the body's heat at rest. **Core** and **shell** temperatures are recognized. During exercise, the amount of heat to be transported to the body surface must increase by tenfold or more. Even in the core, the temperature is not uniform: some organs generate more heat than others, those near veins are cooler than those near arteries, etc. **Deep rectal temperature** is often used as a representative core temperature.

Diurnal temperature fluctuations occur, of about 1–2ºC: the peak temperature is when the animal is most active (e.g. at night for nocturnal species). Such circadian (daily) rhythms are independent of exercise, occurring even in resting animals, and seem to be related to the photoperiod: the cycle is reversible by reversing the illumination cycle [i.e. a nocturnal species kept in a reversed light cycle will exhibit a peak temperature during the day, when it is artificially 'night' (darkness)].

Core temperatures:

- Monotremata (Prototheria) $31 \pm 2ºC$
- Marsupialia (Metatheria) $36 \pm 2ºC$
- Placentalia (Eutheria) $38 \pm 2ºC$
- Aves $40 \pm 2ºC$

'Primitive' orders of placentals (e.g. Insectivora) tend to have a lower core temperatures than do 'advanced' orders (e.g. Carnivora).

Control seems to be sophisticated: related species of mammals from polar and tropical regions have very similar core temperatures; the same applies in birds. The core temperature of sheep, recorded using a telemetric thermocouple, is the same ($\pm 0.1ºC$) whatever the weather at a given time of day, and only varies by $\pm 1.9ºC$ during a whole year.

Regulation of body temperature is mainly under autonomic (parasympathetic) nervous control, although sympathetic fibers control vasodilation, and shivering is under somatic nervous control: temperature detectors are more useful near the peripheries of the body where most variation occurs. **Warm neurons** linked to sensors in the skin increase activity as the temperature rises whereas **cold neurons** increase activity as it falls. Information is also fed in from core temperature sensors. [**Gain** is a useful concept: it measures how far the temperature would shift without negative feedback divided by how far it *does* shift. For human thermoregulation, gain = 33 (i.e. temperature stress which would raise a human corpse from 37–47ºC would actually raise a living, thermoregulating human by 0.3ºC: gain = 10 ÷ 0.3 = 33).]

Heat transfer

If an animal is to maintain constant temperature, **heat loss must equal heat gain**. Thus if heat produced in exercise is 10 times that produced at rest, heat loss must also increase tenfold. Heat loss will relate to air or water temperature, wind or water currents, insulation, etc. Heat flows from higher to lower temperatures through the following mechanisms.

Conduction (direct transfer of kinetic energy of molecular motion between physical bodies in contact with each other); the thermal conductivity coefficient, k, measures ease of heat flow:

k (water) $= 5.8 \times 10^{-3}$ joules sec^{-1} cm^{-1} $^{\circ}$C^{-1}

k (air) $= 2.1 \times 10^{-6}$ joules sec^{-1} cm^{-1} $^{\circ}$C^{-1}

k (fur) $= 3.7 \times 10^{-6}$ joules sec^{-1} cm^{-1} $^{\circ}$C^{-1}

Thus (dry) fur is a poorer conductor (and better insulator) than water. However, these values are complicated by curved surfaces.

Convection is an extension of conduction and involves mass flow of molecules in a fluid; it may be free (due to density changes, warm air rises) or forced (e.g. fanning).

Radiation involves no direct contact between the heat source and its 'sink'; for physiological purposes, air is considered transparent to radiation. The intensity of radiation is proportional to [temperature (in kelvins) of radiating surface]^4; thus emission rises rapidly with temperature.

At physiological temperatures most objects (e.g. skin) emit in the middle infra-red. The emissivity equals the absorptivity if an object is in equilibrium with its environment; thus skin, fur, etc. have high absorptivities in the middle infra-red.

[N.B. heat loss by radiation from pigmented and unpigmented bodies is about the same: color may be important for heat absorption from the sun (e.g. in basking lizards) which has its peak intensity in the visible range.]

Heat loss to cool surroundings = Q.

$$Q = C(T_2 - T_1)$$

where $T_2 - T_1 =$ temperature difference and $C =$ a combined constant reflecting insulation, etc.

$$Q = C(T_B - T_A)$$

where $T_B =$ body temperature and $T_A =$ ambient (environmental) temperature.

Evaporation also uses much heat (through latent heat of evaporation of water ≈ 2400 joules g^{-1}); thus water evaporating from a surface (e.g. skin, tongue) results in loss of a great deal of heat.

Heat balance

For homoiothermy: heat loss = heat gain.

$$H_{total} = \pm H_c \pm H_r \pm H_e \pm H_s$$

where H_{total} is total heat production (positive), H_c is conductive and convective heat loss/gain (positive for net loss), H_r is net radiation heat exchange (positive for net loss), H_e is evaporative heat loss (positive for net loss) and H_s is heat stored (positive if the heat is stored and the body gains heat; temperature rises).

Values to the right of the equation can be negative (e.g. H_e can be negative when a cool body enters a damp atmosphere, e.g. a Turkish bath). In a heat gain situation, a constant temperature can be maintained if evaporative heat loss increases.

Keeping warm

$$H = Q = C(T_B - T_A)$$

where H is heat production and Q is heat loss. The animal must tolerate T_A (the environmental temperature) unless it moves; therefore it must adjust C (conductance) or T_B (body temperature). Normally body temperature stays constant, so heat production or conductance must be altered.

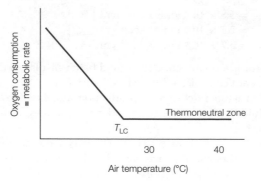

Fig. 1.   The thermal neutral zone of a pigmy possum.

Heat production, *H*, is raised by:

● more exercise (more metabolic heat).
● involuntary muscle contraction (shivering).
● non-shivering thermogenesis [fat in brown adipose tissue (BAT) can be oxidized: it contains a protein which uncouples oxidative phosphorylation from adenosine triphosphate (ATP) production in mitochondria so that energy from oxidation is released as heat].

The metabolic rate stays constant as the air temperature falls and $(T_B - T_A)$ increases: this is the **thermoneutral zone**. Below the **lower critical temperature** (T_{LC}) the metabolic rate rises (*Fig. 1*). T_{LC} has lower values in well-insulated, polar species, and the gradients of the slopes showing the increase in metabolic rate with decreasing temperature are steeper in poorly insulated species.

Conductance measures heat flow from an animal to its surroundings. **Insulation** is the reciprocal of conductance; effective insultation minimizes heat loss. Heat loss is also reduced if the **surface area:volume ratio** is smaller. Larger animals conserve heat more effectively than do small ones with similar body architectures, and rolling into a ball or huddling can effectively reduce surface areas.

Fur is the major barrier to heat flow in land mammals: its insulation is very variable. Insulation increases with fur thickness, reaching a maximum in animals such as arctic foxes. Smaller animals have shorter, lighter fur and tend to be fossorial (burrowers) or hibernators. Polar bears have open, coarse fur which poorly insulates and, in water, most insulation is lost (loss in agitated water is 50 times that in air) so they use subcutaneous **blubber** (fat), as do marine mammals. In many species, winter fur is much heavier than summer fur: a black bear's summer fur has only 52% of the insulation capacity of winter fur.

Many **marine mammals** and birds (e.g. seals and penguins) inhabit polar seas. Water has a high conductance. In the $H = C(T_B - T_A)$ equation, *C* or *H* is altered. The metabolic rate in some marine species is higher than expected: in a harp seal, the metabolic rate is constant down to the freezing point of water. Insulation is by subcutaneous blubber. A countercurrent heat exchanger is found in flippers which lack blubber: returning venous blood is warmed by outgoing arterial blood. Each artery is surrounded by many veins; summer overheating is avoided by a venous shunt (the artery expands and compresses the veins) so that returning blood passes along alternative superficial veins and is not warmed by the arterial blood. Such countercurrent heat exchangers are used in

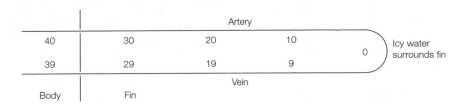

Fig. 2. The structure of a countercurrent heat exchanger in the fin of a polar seal (°C).

terrestrial homoiotherms: they allow a heron to stand in icy water, and are also found in human arms and legs (*Fig. 2*).

Birds, with **insulating down feathers**, do not compensate completely by increasing metabolic rate with declining temperatures, that is in the $H = C(T_B - T_A)$ equation, C (conductance) may drop with falling temperature. Other strategies include raising feathers and drawing the feet up to assume a ball-shape, allowing the shell temperature to drop while the core remains the same.

Keeping cool   Above the lower critical temperature T_{LC}, metabolic heat production stays about constant throughout the thermoneutral zone. In the $H = C(T_B - T_A)$ equation, if H and T_B stay the same, then if T_A rises, C must alter. Heat flow is helped by vasodilation of cutaneous blood vessels and by increasing the surface area. As the environmental temperature goes up, the conditions for losing heat by conduction, convection and radiation become less favorable; evaporative heat loss and heat storage are more emphasized. For example, a human in a desert has a metabolic rate of 293 kJ h^{-1} ($\equiv$ 0.12 liters of water evaporation); in the desert, a human may sweat 1–1.5 liters per hour (e.g. 1.32 l h^{-1}).

If there is no change in T_B, then 1.2 liters of water (1.32 – 0.12) evaporated must be equivalent to the heat gain from the environment: thus the heat load from the environment is 10 times that from metabolism. If T_B can be raised by 1°C, then heat stored is 250 kJ, saving 0.1 liters of evaporated water.

Surface area:volume ratio is again important, as heat gain from the environment is a surface process. Enviromental heat load is related to surface area while metabolic heat production is related to volume. Thus, in arid conditions such as hot deserts, a small animal is at a disadvantage with respect to heat loads. It is possible to predict how much water animals will need to evaporate in order to dissipate heat (*Fig. 3*): a jerboa will need to evaporate 14% of its body weight per hour, a rabbit 7% and a camel 1%; that is an exponential function of body weight. Thus small animals tend to be fossorial (burrowing) and/or nocturnal (e.g. jerboa); the desert jackrabbit shelters in shadows during the day and vasodilates the extensive veins in its ears to lose heat. The camel derives advantages from its large size (and low area:volume ratio). If it thermoregulated in the same way as humans do it would lose about half of the percentage of water per hour, relative to body weight, that a human would. However, it also increases H_s. A dehydrated camel has great diurnal T_B fluctuations, from 34 to 41°C.

For a 500 kg camel, a 7°C T_B rise $\equiv$ 12.1 MJ $\equiv$ 5 liters of water

At night the camel loses heat by radiation; $(T_B - T_A)$ is also reduced as the camel heats up, and this saves water because less is used to defend T_B by

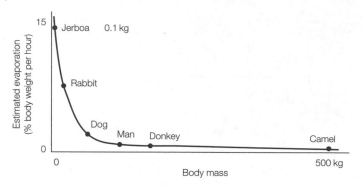

Fig. 3.   *Body size in relation to estimated evaporation rate in desert mammals.*

evaporative cooling. The camel is also helped by light-colored, insulating fur [a shaved camel (!) would evaporate 50% more water]. Camels also tolerate more dehydration (10–12% body water loss is fatal in humans; 24% body water loss is tolerated by camels).

Ground squirrels store heat: they dart around and heat up quickly (large area:volume ratio) but cool off quickly in their burrows.

Sweating and panting are methods used to evaporate water and lose latent heat of evaporation:

- **Sweating** is useful for naked mammals (e.g. humans) or in very dry atmospheres (e.g. camels); sweating is more passive than panting but results in salt loss.
- Ungulates (hoofed mammals) and carnivores tend to **pant**. Many desert animals have long respiratory tracts which act as heat exchangers and water conservation channels – ingoing air is warmed and moistened in the nasal tract, outgoing air is cooled and water condenses. When cooling is required, the outgoing air is diverted over the moist buccal (mouth) membranes from which water is evaporated. Panting can lead to respiratory alkalosis and requires work – therefore, use is made of shallower respiration at greater frequency, using dead space in the upper respiratory tract – a dog panting shifts from 30 to 40 breaths per minute to 300 to 400 breaths per minute, the natural oscillation frequency of the respiratory system. Panting is best under high and sudden heat stress (e.g. a pursued gazelle): the core temperature gets very high but the brain stays cool, the carotid artery passing through a venous sinus near the nasal cavities.
- Rats **salivate over their limbs** in thermal emergencies.
- Pigs and hippopotami **wallow** in mud: the damp mud clings to the body and evaporates slowly – this technique uses external water.

Thermoregulation in poikilotherms

In most poikilotherms $T_A = T_B$. Small aquatic animals lose heat rapidly because of the high conductivity of water, and they cannot have too high a metabolic rate because of oxygen demands – huge gills would be required and these would be sources of heat loss unless a heat-exchanger was built in.

Many large, active fishes (e.g. bluefin tuna) isolate their swimming muscles, and use a countercurrent heat exchanger to conserve metabolic heat, that is they are partial endotherms. Cutaneous arteries supply the swimming muscles. Parallel fine vessels enter the muscle, enmeshed in veins leading out to the cutaneous vein; veins warm arterial blood as it enters. The interior of the muscle can be more than 14°C above the water temperature. High temperatures increase muscle power output, enabling them to contract faster. Other heat exchangers isolate the tuna heart, liver, etc.

On land, radiation and evaporative heat loss and gain are important. To raise the temperature it is important to maximize gain from radiation and storage and minimize conductive and convective loss. Many **lizards** bask in the sun, gaining radiant solar heat and absorbing conductive heat from hot rocks. *Liolæmus* sp., in morning sun at –2°C, raises its cloacal temperature to 33°C within 60 minutes (at an ambient temperature of 1.5°C). This preferred (eccritic) temperature of 33°C is maintained by moving in and out of the sun, orienting the body parallel or at 90° to the sun's rays, lifting the body (or two legs) off the rocks. The phenomenon is known as **heliothermy** and is an example of **behavioral thermoregulation**. Some species alter the color of their skin, making it darker in order to absorb more incident solar radiation in the visible wavelengths. In the marine iguana, cooling when swimming in a cold sea is much slower than heating, due to changes in skin circulation. The lizard *Sauromalus obesus* pants when hot. Because the carotid arteries pass near the pharynx wall panting is adequate to cool the brain, but not the whole body. Brooding pythons 'shiver' to generate metabolic heat to warm their eggs.

Thermal inertia may maintain large reptiles (e.g. monitor lizards, and perhaps applied in Mesozoic 'dinosaurs') at a more or less constant body temperature. The small area:volume ratio means that such animals heat up only a little by day and cool only a little by night, and metabolic heat may not escape (*Fig. 4*).

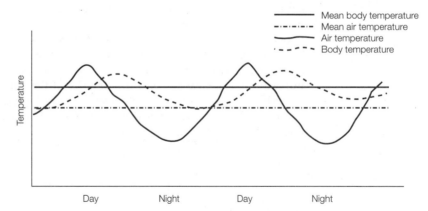

Fig. 4.   'Homiothermy' due to thermal inertia in large reptiles.

Some flying insects warm their flight muscles by shivering. The sphinx moth *Manduca sexta* flies on cool nights with muscles at 35°C. The moth shivers and the wings lightly vibrate due to spontaneous contraction of upstroke and downstroke muscles. Thoracic heat is retained by furry scales; overheating is prevented by circulating hemolymph through the abdomen.

Torpor and hibernation

Keeping the core temperature (T_B) high is expensive, especially in small animals with high metabolic rates. Therefore T_B could be allowed to drop. Many small mammals (e.g. dormouse, hedgehog) and a few birds **hibernate over winter;**

other small animals undergo periods of **diurnal torpor** (e.g. bats, humming-birds). During hibernation/torpor the metabolic rate falls and the animal becomes **torpid** and less responsive to external stimuli. (Torpor is confined to small homoiotherms: bears have a deep winter sleep but T_B is little depressed.) **Estivation** is a period of summer or drought inactivity. For example, Columbian ground squirrels disappear into burrows from August to May. Some snails and lungfishes radically reduce their metabolic activities during droughts.

Torpor is characterized by lowered metabolic rates (oxygen consumption can be 2.5% of normal) and extreme bradycardia, including periods of asystole (cessation of heart beat): this can lead to acidosis. Hypoxia is tolerated because of high hemoglobin levels and a leftward shift of the oxygen–hemoglobin disso-ciation curve. Endocrine function is quiescent, although gonads may become active towards the end of hibernation.

Hibernators feed before hibernation. The phenomenon is triggered by the time of the year, related to endocrine cycles partly supplemented by environmental cues such as photoperiod, temperature and/or availability of food supplies

Hibernation and torpor are under close control: the analogy is with turning down the thermostat in a heating system. The **thermostat is the hypothalamus**: thermoreceptors input to the hypothalamus and appropriate heat-generating or heat-loss mechanisms are elicited. Sensors can be bypassed experimentally by locally heating or cooling the hypothalamus using thermoprobes: the tempera-ture at which no response is elicited is the 'set-point'.

There are three types of **arousal:**

- Alarm arousal (external stimulation), for example low temperature: a hiber-nating hedgehog maintains T_B at 5ºC. Lowered temperatures result in *either* increased metabolic rate *or* arousal.
- Periodic arousal (internal stimulation), for example lowered glucose levels.
- Final arousal: the rate is related to the species, varying from 30 minutes to many hours. It is noted that near arousal time, the temperature set-point fluctuates (for unclear reasons); the anterior of the animal warms before the posterior. In many species, both white adipose tissue and brown adipose tissue (BAT: see above, strategically situated between shoulder blades near the heart) aid warming.

C22 HORMONES

Key Notes

Hormones and nerves	Hormonal control is largely concerned with metabolism, growth and reproduction; nervous control is more concerned with immediate responses. There is considerable co-operation between the two systems.
Hormones and the endocrine system	Hormones are chemical messengers, secreted mainly by endocrine glands, sometimes by neurons. Protein hormones bind to membrane receptors on target cells, activating a second messenger in the cell. Steroid hormones mostly cross the target cell membrane, and bind to a cytoplasmic receptor.
Hierarchy of control	A hierarchy of control is often seen: for example the hypothalamus controls the pituitary which controls the endocrine gland.
Invertebrate hormones: crustacean molt	The crustacean molt is controlled by juvenile hormone from the X gland/sinus gland and ecdysone from the Y gland. This is stimulated by an increasing photoperiod and is an example of invertebrate hormonal control.
Vertebrate hormones	Neurosecretion is pivotal in the vertebrate endocrine system. The hypothalamus affects the pituitary gland which has important controlling effects.
Vertebrate posterior pituitary	In mammals, the posterior pituitary releases neurosecretions from the hypothalamus, including antidiuretic hormone (ADH; vasopressin) (water retention and vasoconstriction) and oxytocin (milk ejection, uterus contraction).
Vertebrate anterior pituitary	The anterior pituitary synthesizes prolactin and melanophore-stimulating hormone, and hormones affecting hormone release from other glands: growth hormone, gonadotropins (follicle-stimulating hormone and leutinizing hormone), thyroid-stimulating hormone, adrenocorticotropic hormone. Most of the anterior pituitary hormones are under the control of hypothalamic releasing factors, and may be influenced by further nervous input and by external factors, for example photoperiod.
Vertebrate hormonal systems	Important vertebrate hormonal systems include: <ul><li>Growth hormone (growth).</li><li>Thyroid hormones (respiratory enzyme synthesis; control of metabolic rate; metamorphosis in Amphibia).</li><li>Melatonin (color change; reproductive cycle control).</li><li>Erythropoietin (red blood cell synthesis).</li><li>Gastrin, secretin, CCK–PZ (control of digestion).</li><li>Insulin, glucagon, somatostatin (control of blood sugar levels).</li><li>Adrenaline (epinephrine), corticosteroids (stress management; fright/fight/flight responses).</li></ul>

- Parathyroid hormone and calcitonin (blood calcium control).
- Prolactin, oxytocin (mammalian birth and lactation; prolactin prevents metamorphosis in Amphibia).
- Reproductive hormones, for example gonadotropins and sex steroids (gametogenesis, egg production, pregnancy, secondary sexual characteristics).

Related topics

Homeostasis (C1)	Metamorphosis (D4)
Blood glucose control (C19)	Puberty in humans (D5)
Integration and control: nerves (C23)	Menstrual cycle in women (D6)
Lactation (D3)	

Hormones and nerves

Increasing size and complexity in multicellular animals necessites comunication systems for integration and control. **Two principal methods** are used:

- **Chemical** control, using chemical messengers or **hormones** within the endocrine system (or, more locally, between cells).
- **Electrical** control, using **nerves**.

Generally, nerves are used for immediate, short-term responses and control, whereas hormones are used for long-term control associated with metabolism, growth and reproduction. There is frequent co-operation and interreliance between the two arms of the integration and control system.

Hormones and the endocrine system

Hormones can be defined as chemical messengers. They are produced by endocrine glands of epithelial origin and are released directly into body fluids such as hemolymph or blood. Hormones are synthesized intracellularly where they may be stored in vesicles, and are released by exocytosis.

Hormones may be:

- water-soluble peptides or proteins (e.g. insulin);
- water-insoluble steroids (e.g. estrogen);
- amino acid derivatives such as catecholamines [e.g. adrenaline (epinephrine)].

Protein, peptide and catecholamine hormones are transported around the body in solution. They bind to receptors on the cell membrane; the hormone–receptor complex activates the production and release of intracellular **second messengers**. (The extracellular protein hormone is the first messenger.) An example of a second messenger is cAMP. Here the protein hormone binds to a receptor and the complex activates adenyl cyclase which facilitates the conversion of ATP to cAMP. cAMP, the second messenger, in turn promotes the activation of protein kinases within the cell which then activate phosphorylases.

Steroid hormones diffuse across the cell membrane where they bind to a cytoplasmic receptor. The hormone–receptor complex is received by an acceptor on the chromosome, resulting in increased transcription of DNA.

Concentrations of hormones in body fluids are low (10^{-11}–10^{-12} M) and are normally under the control of negative feedback processes (see Topic C1). Positive feedback may be utilized; for example high levels of estrogen from the ovary stimulate an increased release of luteotrophic hormone from the anterior pituitary, leading to ovulation, in the menstrual cycle in women (see Topic D6).

Hierarchy of control

There is often a **hierarchy of endocrine control**. Blood levels of thyroxine are under the control of the anterior pituitary through the mediation of thyroid-stimulating hormone (TSH); the anterior pituitary is, in turn, under the control of tropic hormones secreted by neurosecretory cells in the hypothalamus, here secreting thyrotropin-releasing factor (TRF), which stimulates synthesis/release of TSH. High levels of thyroxine inhibit TRF production through a long negative feedback loop and inhibit TSH through a short negative feedback loop (*Fig. 1*).

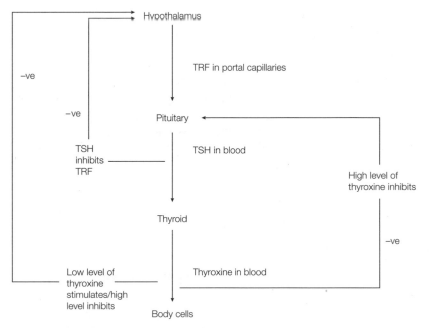

Fig. 1.   Hormonal control of thyroid activity.

An example of **nervous–endocrine co-operation** is seen in milk release in a suckling baby: suckling results in nervous impulses being sent to the brain, which results in increased release of oxytocin by the hypothalamus/posterior pituitary system; oxytocin facilitates milk ejection (see *Fig. 2* of Topic D3).

Invertebrate hormones: crustacean molt

Neurosecretory cells and hormones are found in most metazoan phyla, from Cnidaria to Chordata, the hormones usually affecting reproduction and growth.

Molting during crustacean growth is an example of an invertebrate hormonally controlled system. The epithelially derived Y organ at the eyestalk base produces ecdysone which promotes molting. It is regulated by neurosecretory cells in the X organ of the eyestalk: hormones from the X gland are transported along the axons and are stored in their swollen ends within the sinus gland, also within the eyestalk, and from the sinus gland they are released into the hemolymph. The X gland/sinus gland hormone inhibits molting by preventing release of ecdysone from the Y gland. Increasing day length in the spring results in a nervous inhibition of X gland/sinus gland activity, permitting the release

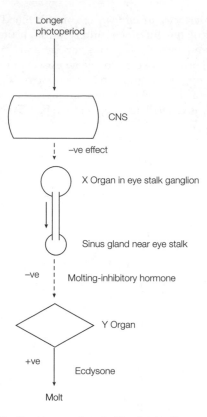

Fig. 2.   Hormonal control involved in the crustacean molt.

of ecdysone and a subsequent molt (*Fig. 2*). (A similar system promotes meta-morphosis in holometabolous insects; see Topic D4.)

Vertebrate hormones

Neurosecretion is pivotal in the vertebrate endocrine system because hypo-thalamic cells affect pituitary gland activity.

Pituitary development
The infundibulum develops into the neurohypophysis (**posterior pituitary**) which connects with that part of the brain (ventral diencephalon) which becomes the hypothalamus. Rathke's pouch develops into the adenohypophysis (**anterior pituitary**); a **portal system** of capillaries and venules links the hypo-thalamus and the anterior pituitary.

Vertebrate posterior pituitary

Posterior pituitary hormones are neurosecretions from hypothalamic cells. ADH (vasopressin) promotes water resorption from the kidney, water up-take across amphibian skin and contraction of arteriolar muscles. **Oxytocin** promotes smooth muscle contraction (e.g. in milk ejection or in the uterus during labor).

Vertebrate anterior pituitary

Anterior pituitary peptide or protein hormones made by discrete cell types include:

- **prolactin** (PRL), controlling production and secretion of mammalian milk, avian nest-building and the maintenance of the larval state in amphibians;
- **melanophore-stimulating hormone** (MSH) which causes the skins of poikilothermic vertebrates to darken by dispersing melanin-containing granules through melanophore cells.

Other hormones affect production and release of hormones by other endocrine glands, giving the anterior pituitary the name 'master gland':

- **growth hormone** (GH) promotes body growth;
- **gonadotropins, follicle-stimulating hormone** (FSH) and **luteinizing hormone** (LH), promote sex steroid synthesis and release by the gonads;
- **thyroid-stimulating hormone** (TSH) stimulates the production of thyroid hormones;
- **adrenocorticotropic hormone** (ACTH) stimulates corticosteroid release from the adrenal cortex.

Many anterior pituitary hormones are themselves under the control of **releasing hormones** from the hypothalamus: hormones such as **gonadotropin-releasing factor** (or hormone) are synthesized by neurosecretory hypothalamic neurons and reach the pituitary via the pituitary portal capillaries. Nervous input and environmental cues such as day length also influence endocrine secretion. GH and PRL secretion are suppressed by hypothalamic inhibiting hormones.

Vertebrate hormonal systems

Among the systems in vertebrates (many described elsewhere in this book) are:

- **GH** produced by the anterior pituitary; this works with somatomedin from the liver and thyroid hormones to promote and control growth. GH promotes proliferation of cartilage in epiphyseal plates of bones and the growth of connective tissues. It also promotes somatomedin synthesis which encourages cell division and differentiation. Low GH levels induce hypothalamic GH-releasing factor; high levels induce GH-inhibiting factor (somatostatin). Excess GH leads to acromegaly (excessive bone development leading to facial deformities and gigantism).
- **Thyroid hormones** including thyroxine and tri-iodothyronine, synthesized/secreted under the influence of anterior pituitary TSH. Thyroid hormones activate genes, promoting, particularly, respiratory enzyme synthesis. In amphibians, thyroid hormones promote metamorphosis.
- **Melatonin and the pineal gland**: melatonin is secreted in darkness (the pineal has a neural link to the eye) and influences color changes, breeding cycles, etc. Melatonin inhibits ovary and testis development in rodents. (In many poikilotherms, e.g. *Sphenodon punctatum*, the tuatara, a primitive lizard-like reptile, the pineal develops into a photoreceptive organ, a 'third eye'.)
- **Erythropoietin** from the kidney promoting erythropoiesis; a number of thymic factors (e.g. **thymosin**) are involved in T-lymphocyte development (see Topics C12–C14). Other hormone-like substances which promote growth are known, including cytokines such as interleukins involved in control of

immune responses, a nerve growth factor involved in neuron growth and putative **thrombopoietin** involved in platelet production.

- **Gastrin, secretin and CCK–PZ**, secreted by the alimentary tract, involved in control of digestive processes (see Topic C18).
- **Insulin and glucagon**, from the pancreas involved in the homeostatic control of blood glucose levels (see Topic C19).
- **Adrenaline (epinephrine)**, formed by methylating noradrenaline (norepinephrine) through the mediation of glucocorticoids, produced by the chromaffin cells of the adrenal medulla; it reinforces the effects of the sympathetic nervous system in stress and fright/fight/flight responses.
- **Glucocorticoids** (e.g. cortisol) secreted by the adrenal cortex; they are involved in protein and carbohydrate metabolism and, with adrenaline (epinephrine) are associated with stress control. They suppress many inflammatory and immune responses. Also produced in the adrenal cortex are cortical androgens (important in women for some muscle and body hair development – in men testosterone from the testes is more important) and aldosterone, important for electrolyte balance and blood pressure control (see Topics C16 and C17).
- **Parathyroid hormone** secreted by the parathyroid glands in response to lowered blood calcium levels: it facilitates calcium uptake from bone and its retention by the kidneys. Vitamin D (cholecalciferol) is made in the skin when it is exposed to sunlight: it increases calcium and phosphate uptake from the gut.
- **Reproductive hormones** including gonadotropins from the anterior pituitary which stimulate the production of **sex steroids from the gonads**. These are involved in gametogenesis, ovulation and reproductive cycles, such as the menstrual cycle in women, human puberty and menopause, and the development of secondary sexual characteristics.

C23 INTEGRATION AND CONTROL: NERVES

Key Notes

Neurons
Neurons comprise cells bodies, dendrites and axons. In higher animals, Schwann cells form a myelin sheath, wrapping round the cells many times in a myelinated fiber. The sheath is interrupted by nodes of Ranvier.

Resting potential
Neurons have a resting potential across the cell membrane; cells are negative inside, and have a high sodium charge outside.

Action potential
An action potential occurs when a stimulus attains a threshold strength, sodium gates open, sodium ions rush in, and the membrane depolarizes – membrane potential rises from –70 mV to +40 mV. Then the potassium gates open and the sodium gates close, which restores the resting potential by efflux of potassium. There is a refractory period after the action potential before the resting potential is re-established. Local circuits facilitate propagation of the impulse along the neuronal cell membrane.

Language and speed
The language of the system is the impulse frequency. The speed of impulse transmission is proportional to the diameter of the axon. High speed transmission is facilitated by saltatory propagation of the impulse at the nodes of Ranvier.

Synapses
Synapses are junctions between neurons, and act as control switches.

Electrical synapses
Electrical synapses (e.g. in the CNS) have gap junctions permitting ion transmission.

Chemical synapses
Chemical synapses use neurotransmitter chemicals. These neurotransmitters are in vesicles in presynaptic neurons. The arrival of the action potential allows voltage-dependent calcium ion channels to open; the vesicle fuses with the cell membrane and neurotransmitter flows into the synaptic cleft. The neurotransmitter binds to the gated receptors on the postsynaptic cell membrane. Sodium flows in and the membrane depolarizes, leading to an action potential. The transmitter is resorbed or broken down.

Excitation and inhibition
One or several presynaptic impulses trigger an excitatory postsynaptic impulse. Other transmitters may bind to postsynaptic receptors, open the potassium gates and hyperpolarize the membrane making it less excitable. This is an inhibitory effect.

Vertebrate nervous system organization
The vertebrate nervous system comprises the central nervous system (CNS: brain and spinal cord) and the peripheral nervous system. Sensory neurons connect sense organs to the CNS; to interneurons in the CNS; and

motor neurons connect the CNS to effector organs. The somatic nervous system allows voluntary actions. The autonomic nervous system controls involuntary actions and is divisible into the parasympathetic (e.g. digestion) and sympathetic (e.g. fright/fight/flight) arms.

Control and reflex arcs	The nervous system obtains, sifts and acts on information. A simple reflex stimulus elicits the appropriate motor response. Modification is possible by countermanding CNS activity.
Evolutionary aspects	Simple animals have nerve nets. Most bilaterally symmetrical Metazoa have a CNS and peripheral nerves. Arthropods have a double ventral nerve cord with segmental ganglia. Chordates have a hollow dorsal nerve cord.
Related topics	Hormones (C22) Sense organs (C25) Integration and control: brain (C24)

Neurons

The nervous system connects **receptors** (such as light or chemical receptors) to **effectors** (such as muscles and glands). The connection is rapid through a network of **nerves**. The nerve cell is a **neuron** (*Fig. 1*), the structural and functional unit of the nervous system. In most complex animals, the nervous system can be divided into the **central nervous system** (**CNS**, comprising the **brain** and **spinal cord** in vertebrates), and the **peripheral nervous system**.

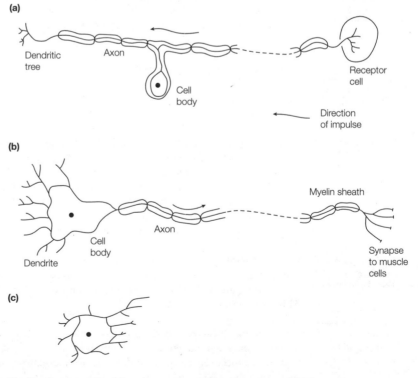

Fig. 1. A mammalian (a) sensory neuron; (b) motor neuron; (c) interneuron.

Neurons have a cell body with a nucleus, and organelles for metabolism. The receptive parts of neurons are short processes, **dendrites**, although information may be received by the cell body. **Axons** are long processes, normally one per neuron, involved in transmitting messages to the distal end where there are small, branching dendrites which facilitate transfer of the impulse across the **synapse** (connection or function) to the next neuron.

Unipolar neurons have a cell body set to one side of a long axon (e.g. vertebrate sensory neurons, *Fig. 1a*); **multipolar neurons** have numerous dendrites entering the cell body and one axon emerging from it (e.g. vertebrate motor neurons; *Fig. 1b*). **Bipolar neurons** have an axon leading to the cell body and one leading away from it (e.g. invertebrate sensory neurons, or the bipolar neurons of the vertebrate eye).

Axons in Cnidaria are naked, but most axons in higher animals are sheathed in **Schwann (neurilemma) cells**. In unmyelinated nerves, Schwann cells enricle the axon once; in myelinated nerves the Schwann cells wrap round the axon many times to form a fatty **myelin sheath**: this sheath is interrupted every 1–2 mm by **nodes of Ranvier**. Here, successive Schwann cells, aligned along the axon, come together but leave a tiny gap, exposing the axonic plasma membrane. Neurons are interspersed by packing **neuroglial cells**.

Resting potential

The plasma membrane of the nerve cell has a **resting (electrical) potential** of −70 mV across it. The inside of the membrane is negative due to negatively charged proteins and PO_4^{3-} inside, giving an unequal distribution of charge. There is a high K^+ charge inside and a high Na^+ charge outside; these ions can (with difficulty) diffuse in and out through sodium and potassium trans-membrane protein channels. K^+ leaks out more easily, enhancing the positive charge outside. Other protein complexes form Na^+/K^+-adenosine triphosphatase (ATPase) pumps which pump sodium and potassium out in a ratio of about 3:2 (*Fig. 2*).

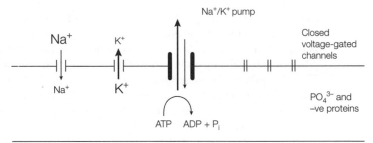

K+ leaks out
faster than
Na+ leaks in

Fig. 2. The axon resting potential.

Action potential

The plasma membrane has proteins that form sodium and potassium voltage-gated channels: voltage changes across the membranes alter the shapes of the proteins, opening and closing the channels. The channels are closed at rest. When the axon is activated, the sodium gates open, sodium ions flow in and the membrane potential rises to about 0 mV; these changes result in the sodium

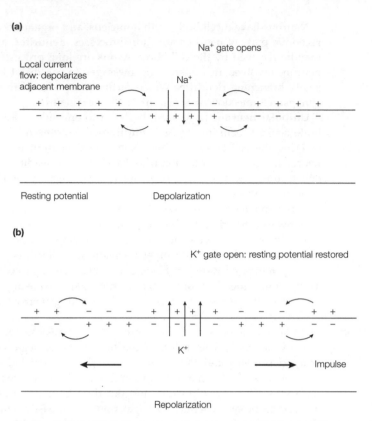

Fig. 3.   *Action potential in a mammalian nerve (in an intact animal the properties of the synapse mean that the impulse only travels in one direction). (a) Activation; (b) recovery.*

gates opening widely, sodium flooding in and the membrane rapidly depolarizing to give a membrane potential of +40 mV, the positive charge on the inside. The gates are open for less than 0.0002 seconds before closing (*Fig. 3a*).

Depolarization only occurs when the stimulus attains a threshold strength: then the action potential is generated in an all-or-nothing fashion. The action potential in a given neuron is the same regardless of the intensity of the stimulus (above threshold strength) or the type of information (e.g. smell, touch, activation of a muscle). During the action potential, the potassium gates open slowly; when the sodium gates close the potassium gates open fully, potassium leaves and the resting potential is restored. The whole process takes about 0.0005 seconds (*Fig. 3b*).

Propagation of impulse

A local current between the depolarized and polarized regions of the membrane activates adjacent parts of the membrane, sodium gates open, depolarization occurs, and so on. Since the impulse starts at the dendrite/cell body end of the cell, it normally proceeds from it to the other end of the neuron (if the axon is stimulated in the middle the impulse will travel in both directions). In nature, the impulses cross synapses in one direction, from the axon of the presynaptic cells to the dendrite of the postsynaptic cell. The impulse depends on local currents generated as a result of the redistribution of ions: thus it does not weaken. The impulse does not reverse itself: sodium gates behind the wave of depolarization cannot re-open until the resting potential is re-established after the efflux of potassium ions: this gives a short **refractory period** of less than a millisecond – thus several thousand impulses per second can be transmitted.

Language and speed

The **language** of the nervous sytem is the **frequency of impulses** (compare Morse code), the frequency being proprotional to the strength of the stimulus above threshold level. The **velocity** of the impulse is proportional to the cross-sectional area of the axon. Giant axons with a diameter of 1 mm are found in invertebrates such as squids, with an impulse velocity of 35 m per second (compare sea-anenome axons with a velocity of <0.1 m per second). Giant unmylenated axons are found in lower vertebrates (e.g. diameter of 50 μm in Amphibia), but in most vertebrates **myelination** increases conduction velocity (e.g. a mammalian axon with a diameter of 10 μm can have a conduction velocity of 120 m per second) and also allows the development of complex nervous systems without the use of excessive space. The myelin acts as an insulator, and action potentials are generated only at the nodes of Ranvier where the axon plasma membrane is exposed. Depolarization at the node leads to a flow of current between it and the next node which depolarizes in turn: this jumping or **saltatory** propagation facilitates great rapidity of conduction (*Fig. 4*). Less ionic pumping saves energy.

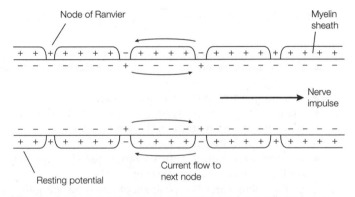

Fig. 4.   Saltatory conduction of an impulse along a myelinated axon.

Synapses

Synapses act as control switches in the nervous system: messages travel from **axon (synaptic) knobs** in the presynaptic neuron to the membrane of the post-synaptic neuron. The synapse permits modification of the message through magnifying it, blocking it, speeding it or slowing it.

Electrical synapses

In **electrical synapses** the two cells are intimately associated at gap junctions which have tiny passages through which ions pass. They allow rapid trans-mission, especially between giant neurons in invertebrates or in the vertebrate CNS.

Chemical synapses

In **chemical synapses** the enlarged presynaptic axon knobs closely appress the dendrite or cell body of the postsynaptic neuron, but a 20 nm **synaptic cleft** separates them. Axon knobs contain 50 μm synaptic vesicles filled with **neuro-transmitter** (e.g. acetylcholine) (*Fig. 5*). The process takes place by the following steps:

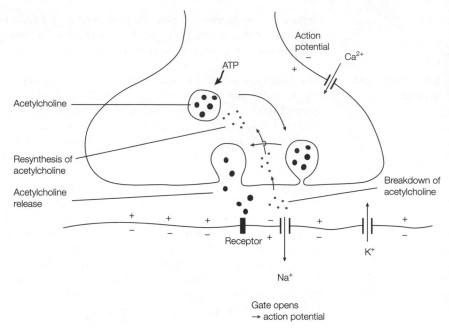

Fig. 5.   A chemical synapse in the mammalian nervous system.

- excitatory nerve impulses reach the axon knob;
- the potential is changed across the plasma membrane of a presynaptic neuron;
- voltage-gated calcium channels open;
- calcium influx causes synaptic vesicles to fuse with the plasma membrane of a presynaptic neuron; ATP provides energy;
- a neurotransmitter is released;
- the neurotransmitter diffuses across the cleft and binds to gated receptors on the postsynaptic membrane;
- Sodium ions enter the postsynaptic neuron via gates, the membrane depolarizes, and the action potential is initiated;
- an enzyme (e.g. acetylcholinesterase) (present on the synapse membrane) breaks down the neurotransmitter, the potassium gates open and the resting potential is restored in the postsynaptic neuron (some neurotransmitters are not degraded but are pumped back intact into the presynaptic neuron);
- neurotransmitter breakdown products are recycled after diffusing back into the presynaptic neuron axon knobs.

Excitation and inhibition

Many presynaptic neurons may synapse with the dendrites and cell body of a postsynaptic neuron. An excitatory impulse from a presynaptic neuron causes an impulse within a postsynaptic neuron: one such presynaptic impulse may not reach the threshold value but further impulses may find it more easy to activate the postsynaptic neuron provided that they arrive before the first impulse has decayed. Several impulses may have a summative (additive) effect, leading to the generation of a postsynaptic neuronal impulse if they occur very closely together in time (temporal summation) or in space on the postsynaptic cell membrane (spatial summation).

Some transmitters, from different presynaptic neurons, bind to receptors on postsynaptic neurons and open potassium gates so that potassium ions flow out, leading to membrane hyperpolarization. This causes the postsynaptic neuron to be less excitable; that is postsynaptic inhibition.

Vertebrate nervous system organization

As noted above, in vertebrates the nervous system can be divided into **brain** and **spinal cord** (the **CNS**) and **peripheral nerves**. **Sensory neurons** (*Fig. 1a*) in sensory nerves convey messages from receptors to the CNS. **Motor neurons** (*Fig. 1b*) in motor nerves convey messages from **interneurons** (*Fig. 1c*) in the CNS to effectors such as muscle or gland cells. The cell bodies for sensory and motor neurons are situated in the dorsal root and ventral horn of the spinal cord respectively, and are not strictly in the nerves. The brain consists of millions of interneurons grouped into **centers** connected by **tracts**. Some centers are concerned with senses (e.g. visual), others are connected with different responses (motor areas), and still others are correlation and co-ordination areas. The spinal cord has ascending, dorsal tracts connecting sensory areas of the body with sensory areas of the brain, ventral descending tracts connecting motor areas of the brain with effectors, and interneurons.

Physiologically the nervous system can be divided into:

(1) **Somatic** (voluntary or skeletal) **nervous system**, concerned with external situations and so with voluntary actions. In humans, the neurotransmitter is acetylcholine.
(2) **Autonomic** (involuntary) **nervous system**, concerned with involuntary responses. It contains:
   • **sympathetic neurons** (largely concerned with responses related to fight/flight/fright). In humans the neurotransmitter is noradrenaline (norepinephrine);
   • **parasympathetic neurons** (concerned with digestion, etc.). In humans the neurotransmitter is acetylcholine.

The autonomic system is a motor system. Neurons extend from the CNS to peripheral motor ganglia (e.g. the solar plexus) where they synapse with post-ganglionic motor neurons which continue to effector organs. The sympathetic and parasympathetic systems antagonize each other: most body organs (e.g. heart) are innervated by both. Parasympathetic fibers release acetylcholine as a neurotransmitter whereas sympathetic fibers release noradrenaline (norepinephrine) which has similar effects to those of adrenaline (epinephrine) from the adrenal medulla. Sympathetic effects include an increased rate of and more forceful heartbeat, increased blood pressure and blood flow to muscles, bronchodilation, increased levels of blood sugar, hair erection, sweating and pupil dilation. Blood flow to and peristalsis in the gut is inhibited. Parasympathetic effects are broadly the opposite to the sympathetic effects: more blood flow to the gut whose motility increases, increased enzyme secretion, enhanced insulin release. Parasympathetic stimulation is also associated with sexual activity.

Control and reflex arcs

Reaction to the environment involves obtaining, sifting and acting on information. Nerve impulses are conveyed to rather few cells and their message is specific: precise signaling is necessary for efficient co-ordination in complex animals, for example a **reflex arc**, such as that stimulated by a finger on a pin. It is advantageous to survival to act upon the pain stimulus, which must first be assimilated. The process has three stages:

- receipt of information by sense organ;
- processing of information and selection of appropriate response;
- response (muscular contraction to remove hand from pin).

The sense organ acquires information and translates it into nerve impulses which are transmitted by sensory neurons to the spinal cord in the CNS. Information is relayed to interneurons in the CNS; a response impulse is then passed down motor neurons to muscle cells where it is translated into a muscular contraction. The same stimulus always gives the same response (e.g. a knee-jerk reaction; dilation of the pupil of the eye in dim light). Such a **simple reflex** (*Fig. 6*) develops early and enhances survival. Note that there are many cells in each receptor, many sensory, motor and interneurons, and many muscle cells.

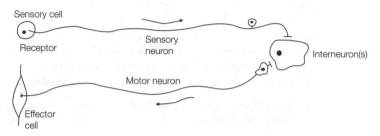

Fig. 6.   A simple reflex arc.

In practice, most reflexes are more complex. Continuous sensations such as touch 'wear off' (**habituation**): receptors respond to change. Habituation causes cessation of receptor signals: the nervous system is not an automaton.

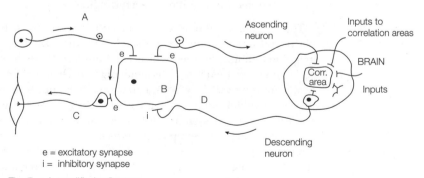

Fig. 7.   A modified reflex arc.

Reflexes may also be modified (*Fig. 7*): we may 'want' to prick ourselves (e.g. for an immunizing injection). Here 'will' overcomes the reflex due to modifications at the interneuron level.

(1) Axons in the spinal tract connect the brain to interneurons.
(2) Messages are conveyed to higher centers of the brain (usually no action is taken, but if other instructions are given, e.g. a nurse asking you to hold your arm still!) countermanding instructions are sent down the spinal cord to prevent muscle contraction.

(3) At the cellular level:
- interneuron B is triggered by impulses from sensory neuron A;
- interneuron B triggers motor neuron C through an excitatory synapse;
- the synapse from the descending neuron axon D releases a different neurotransmitter, leading to hyperpolarization of the membrane of C, preventing the generation of an action potential.

Evolutionary aspects

Sponges lack a nervous system. In Cnidaria (e.g. jellyfishes) an extensive nerve net co-ordinates responses. Increased numbers of neurons and interconnections characterize more complex metazoan animals. Most bilaterally symmetrical animals have a CNS and peripheral nerves.

In annelids and arthropods, there is a ventral (double) nerve cord with segmental ganglia. In the head is a cerebral ganglion ('brain') from which lead circumesophageal commisures which rejoin beneath the gut to form the anterior end of the ventral nerve cord. Chordates are characterized by a hollow dorsal nerve cord formed by the infolding of a neural streak along the embryo's back: in most urochordates the dorsal nerve cord degenerates to a small ganglionic structure after metamorphosis. The octopus brain is similar in terms of processing capacity to that of some mammals.

C24 INTEGRATION AND CONTROL: BRAIN

Key Notes

Brain development	The vertebrate brain develops from expansions of the dorsal nerve cord, giving the fore-brain, mid-brain and hind-brain.
Fore-brain	The fore-brain comprises the:

- cerebrum (cerebral hemispheres): used for correlation, integration, memory, learning and intelligence;
- corpus striatum: used for autonomic control;
- hippocampus: used for receipt of sensory impulses; it is also part of the limbic system (with the hypothalamus and corpus striatum) – important in emotions and motivation;
- pineal: important in light reception, and in reproductive cycles;
- hypothalamus: this is the control center for visceral activity and homeostatic control; it is also part of the limbic system (see hippocampus);
- thalamus: the processor for sensory input and motor output; important in locomotion and in posture.

Mid-brain	The mid-brain comprises the optic lobes (tectum), important in optic and auditory reflexes.
Hind-brain	The hind-brain comprises the:

- cerebellum: used in posture, balance and motor co-ordination;
- pons: a relay center;
- medulla oblongata: the lower center control for ventilation, cardiac rhythms, etc.

Neurotransmitters	Defects in levels of neurotransmitters in the brain can lead to neurological and psychiatric disorders. Endorphins and encephalins act as natural opiates.
Spinal cord	The spinal cord continues from the hind-brain. It has dorsal sensory columns with ascending tracts and ventral motor columns with descending tracts, together with interneuronal regions.
Related topics	Hormones (C22) Sense organs (C25)
	Integration and control: nerves (C23)

Brain development

The brain develops as three basic expansions at the anterior end of the hollow dorsal nerve cord. These form the fore-brain, mid-brain and hind-brain respectively (the fore-brain and hind-brain further divide).

• Fore-brain (Prosencephalon)	Telencephalon	Olfactory bulbs
		Cerebral hemispheres or cerebrum
		Hippocampus
		Corpus striatum
	Diencephalon	Pineal body
		Thalamus
		Hypothalamus
		(Neural lobe of pituitary)
		(Retina)
• Mid-brain	Mesencephalon	Optic lobes
• Hind-brain (Rhombencephalon)	Metencephahlon	Cerebellum
		Pons
	Myelencephalon	Medulla oblongata

The cavity of the dorsal nerve cord expands to form the **ventricles** of the brain, filled with **cerebrospinal fluid** (CSF). **Meninges** form a barrier between the brain (and spinal cord) and skull bones (and vertebrae). Dorsal and ventral columns of gray matter (neuron cell bodies, dendrites and unmylenated axons) in the spinal cord continue in the brain to form nuclei associated with cranial nerves into and out of the brain. Dorsal nuclei are concerned with sensory function, ventral with motor function. Gray matter is also found on the surface of the cerebral hemispheres. White matter forms tracts in the brain: myelinated axons and packing cells.

Fore-brain

Cerebrum (cerebral hemispheres)
This has an olfactory function in primitive vertebrates; in higher vertebrates the cerebrum expands and the outer cortex becomes the **main correlation and integration center** of the brain, concerned with learning, instinct, 'intelligence' and memory.

Much of the cortex is made up of association areas which interact with primary sensory and motor areas; sensory stimuli are integrated, interpreted and acted upon through motor impulses. Short-term memory is based on neuronal circuits; long-term memory is probably due to microanatomical and biochemical changes in neurons and synpases, setting up memory traces.

Corpus striatum
This is the center for many autonomic activities.

Hippocampus
This receives sensory impulses; with part of the corpus striatum and hypothalamus it forms a major part of the **limbic system** influencing motivations and emotions concerned with processes such as feeding, drinking, fighting, sleep and reproduction.

Pineal body
In many lower vertebrates, the pineal forms a light-receptive 'eye'; in higher vertebrates, it forms the pineal gland, linked to the optic nerve and secreting melatonin. In many species it influences reproductive cycles.

Hypothalamus
This is the **main center for visceral integration and homeostatic control**. It receives sensory signals, and its own cells respond to glucose, water, electrolyte, hormone levels and blood temperature. It controls sleep, hunger, thirst, digestion, glucose, salt and water levels, body temperature (in homoiotherms) and sexual activity, relating to centers in the cerebrum. The hypothalamus facilitates its effects via the autonomic nervous sytem and the anterior and posterior pituitary gland.

Thalamus
This is a processing center for sensory impulses (except olfactory) and motor impulses concerned with posture and locomotion.

Mid-brain

Optic lobes (tectum)
The optic lobes are a major integration center for sensory input in lower vertebrates. Later in evolution the cerebrum assumes this role, but optic lobes are still important for optic and auditory reflexes.

Hind-brain

Cerebellum
This is the center for **motor co-ordination**, especially relating to balance, posture and locomotion.

Pons
The pons is important as a relay center between the cerebrum and cerebellum.

Medulla oblongata
The medulla connects to the spinal cord; it is the **lower center control** for feeding, swallowing, ventilation and heartbeat; respiratory and cardiac centers have an inherent rhythmicity, modified by sensory inputs. The medulla links, via the **reticular formation**, to higher brain centers.

Neurotransmitters   Neurotransmitters (some excitatory and some inhibitory) in the brain include acetylcholine, noradrenaline (norepinephrine), serotonin, g-aminobutyric acid (GABA) and dopamine. Defects in levels of neurotransmitters can lead to neurological and psychiatric disorders; for example, low levels of dopamine are associated with Parkinson's disease. Endorphins and encephalins are peptides secreted by neurons and they act as natural opiates suppressing pain and (in humans) affecting emotions.

The spinal cord   The spinal cord continues from the hind-brain; it has 31 segments in humans, each giving rise to a pair of spinal nerves. There are dorsal sensory columns containing ascending tracts (funiculi) and ventral motor columns with descending tracts, together with interneuronal regions.

C25 SENSE ORGANS

Key Notes

Sensory receptors	Internal and external stimuli are sensed by sensory receptors, which are often grouped into sense organs. Stimuli include pressure, temperature, chemical substances, vibrations, mechanical deformation and radiant energy.
Receptor potential	The stimulation of a receptor leads to a receptor potential; if this exceeds a threshold, an action potential is initiated leading to a flow of information through the nervous system.
Types of receptor	Receptors can be divided into mechanoreceptors, chemoreceptors and radioreceptors. Further division is possible into interoreceptors, proprioreceptors and exteroreceptors.
Chemoreception	Chemoreception includes taste and smell. The olfactory receptors comprise bipolar neurons whose cilia have receptors with shapes complementary to odor molecules.
Mechanoreceptors	Mechanoreceptors include free nerve endings (often associated with hairs), and Meissner's corpuscles and Merkel's cells in skin. Pacinian corpuscles detect pressure. Lateral line systems with neuromast cells in fishes sense movements in water. Specialized neuromast cells can detect electric currents.
Gravitational pull and balance	Gravitational pull is sensed by statocyst organs containing statoliths. A membranous labyrinth in the vertebrate ear, with semicircular cells, detects gravity and acceleration and provides information concerning position and balance.
Hearing and sound	Sound is detected by sensillae or tympanic ears in insects. Fishes use the lateral line system (and also use the swim-bladder as a hydrophone). Land vertebrates use ears.
The mammalian ear	Mammalian ears collect sound waves which vibrate a tympanum. These vibrations are transmitted via auditory ossicles to the oval window leading to the cochlea. The vibration of fluid in the cochlea leads to movement of the basal membrane and so to vibration of sensory hair cells (= mechanoreceptors).
Proprioception	Proprioreceptors in muscles and tendons signal stretch and contraction.
Photoreception and vision	Photoreceptive cells containing a carotenoid pigment are usually concentrated in the eyes. Light-orientating eyes detect direction of light. Image-forming

eyes use a lens to defract light to form an image. Compound eyes comprise many ommatidia, these often have poor resolution but can detect movement.

Thermoreception   Temperature reception is normally by free or encapsulated nerve endings; in some snakes they are concentrated in pits on the head.

Related topics   Integration and control: nerves (C23)   Integration and control: brain (C24)

Sensory receptors

Information received, processed and transmitted by neurons and synapses of the nervous system is carried to the CNS by sensory neurons (see Topic C23). The triggering of nerve impulses in the sensory neurons depends on translation (transduction): the conversion of one form of energy, the energy of the **stimulus** (e.g. light radiation) to another form of energy, the electrical energy of the nerve impulse.

Stimuli include pressure, heat/cold, chemical substances, vibrations and mechanical deformation, and radiant (electromagnetic) energy. Stimuli are sensed by sensory **receptors**, often grouped in **sense organs** comprising receptors and associated cells.

Receptor potential

When a sensory receptor cell is stimulated, its membrane permeability to ions (or that of a local sensory neuron) is altered and a receptor potential is generated or changes (compare nerve action potential, Topic C23). If the receptor potential reaches a threshold, an action potential is initiated which starts a flow of information through the nervous system. The more intense the stimulus, the greater the frequency of action potentials. Differences between senses lie in their reception and interpretation within the CNS, although the sensory receptors may need to discriminate between different forms of stimuli.

Types of receptor

- Mechanoreceptors: touch, position, hearing, pain
- Chemoreceptors: taste, smell
- Radioreceptors: vision, heat

Electrical current and, arguably, magnetic fields may also be sensed in some animals.

Receptors can also be divided into:

- Interoreceptors
- Proprioreceptors
- Exteroreceptors

Interoreceptors (interoceptors) include mechanoreceptors and chemoreceptors sensitive to blood pressure (baroreceptors), oxygen, carbon dioxide and pH in body fluids, and temperature sensors in the hypothalamus. An animal is not normally conscious of their signals although they may signal pain, hunger, thirst, nausea or (via stretch receptors) a full bladder or bowel.

Proprioreceptors (proprioceptors) provide information about the orientation of the body in space and the position of the limbs, etc. Proprioception enables us to tie our shoelaces in the dark!

Exteroreceptors (exteroceptors) provide information about the outside world; some are very simple – **free nerve endings** sense pain and temperature.

Chemo-reception

Chemoreception includes what are commonly called **taste and smell**, although interoreception of body fluid oxygen, carbon dioxide and pH are also forms of chemoreception.

In vertebrates, **taste** cells are located in the mouth: they act as sentinels to advise on whether food is swallowed or rejected. In humans there are four basic tastes: sweet, sour, salt and bitter; cells, of four types, are grouped in buds with a pore opening to the surface. Most distinctive 'tastes' (gustatory flavors) are smell/taste/tactile combinations.

Smell (olfaction) is chemoreception of airborne or waterborne substances; it is the dominant sense in many species. In mammals, the substances must first be dissolved in a watery mucus overlying the olfactory epithelium in the nasal cavities. In humans the olfactory epithelium covers an area of about 4 cm^2, with about 50 million receptors per side. Receptor cells are bipolar neurons situated in the olfactory regions of the nasal tract. These have cilia whose membranes contain specific receptor molecules into which odor molecules of a complementary shape fit (lock-and-key); binding triggers generation of an action potential. (It is said that humans can distinguish about 10 000 odors made up of seven primary odors: camphoric, musky, floral, pepperminty, ether, pungent and putrid.)

In arthropods, nonmotile olfactory cilia lie in perforated pits or bristles (setae) in or on the exoskeleton. These **sensilla** (*Fig. 1*) are usually on the antennae of insects and are important for detecting odoriferous pheromones produced by female insects.

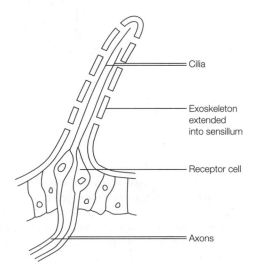

Cilia

Exoskeleton extended into sensillum

Receptor cell

Axons

Fig. 1. An insect olfactory sensillum.

Mechano-receptors

Tactile stimuli such as touch and pain, together with sensing gravitational pull and vibrations in the medium (e.g. sound), are detected by mechanoreceptors. Combinations of a **free nerve ending with a hair** (*Fig. 2a*) supply an exquisitely sensitive mechanoreceptor: bending hair (e.g. in a cat's whisker) triggers, even by air movement, an action potential in sensory neurons.

Other mammalian skin exteroreceptors include **Meissner's corpuscles** (*Fig. 2b*) and **Merkel cells** (*Fig. 2c*) concentrated in sensitive areas of the skin (e.g. human finger tips) giving cutaneous sensitivity. **Pacinian corpuscles** (*Fig. 2d*), deeper in the skin, respond to pressure and vibration: they comprise nerve endings surrounded by layers of tissue and fluid (in an onion formation): pressure deformation of the corpuscle stimulates the sensory neuron. Free nerve endings may also sense pain (*Fig. 2e*).

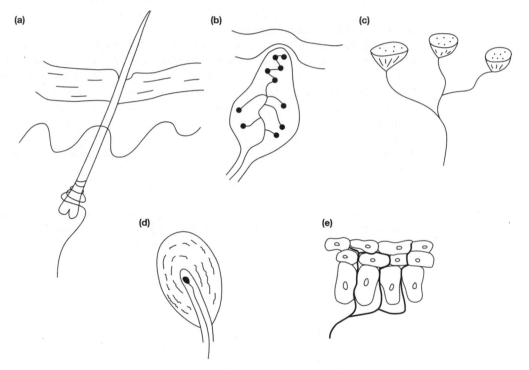

Fig. 2.   *Mammalian skin receptors. (a) A nerve ending associated with a hair. (b) A Meissner's corpuscle (touch sensitive). (c) Merkel cells (touch sensitive). (d) Pacinian corpuscles (sensitive to pressure and vibrations). (e) Free nerve endings (sensitive to pain and temperature).*

Fishes and amphibians have a **lateral line system** (*Fig. 3a*) which allows detection of movements such as currents and vibrations in water. Receptor **neuromast cells** (*Fig 3b*) lie in water-filled canals which communicate to the exterior by pores; the neuromast cells have a single 'hair' (a nonmotile cilium) extending into a cap (cupola). Water movements bending the cup towards the cilium increase the receptor potential and the frequency of impulses; those bending the cup away have the opposite effect. Specialized lateral line receptors in some fishes (e.g. sharks) can detect **electric currents** produced by the muscular activity of other animals.

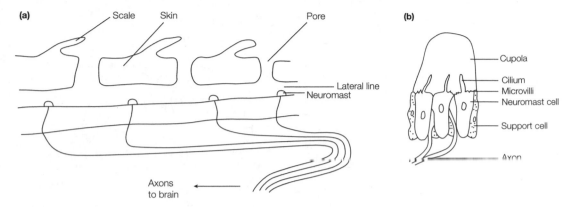

Fig. 3.   *The lateral line of a fish (a) containing neuromast receptor cells (b).*

In insects, **tactile sensilla** comprise a chitinous filament articulating by a thin, flexible joint to the exoskeleton, for example on the antennae; movements of the filament base stimulate an adjacent neuron.

Gravitational pull and balance

Gravitational pull is detected by **statocysts**, thus enabling animals to sense their orientation in space. Statocysts (e.g. in cnidarians, platyhelminths and molluscs) are hollow spheres containing receptor cells with nonmotile cilia. Calcareous **statoliths** are found free in the sphere. The position and movement of the statoliths are detected by the statocyst receptor cells and subsequently interpreted in the CNS as the body's orientation. In decapod crustaceans (e.g. crabs), sand grains imported from outside serve as statoliths in statocyst pits at the bases of the antennae; following a molt the statoliths are lost and new statoliths are obtained as replacements. (Magnetite crystals have been found in some vertebrate cells, inviting the suggestion that they may play a part in reception of information concerning the Earth's **magnetic field**, perhaps of relevance to migration.)

The vertebrate ear (*Fig. 4*) has a **membranous labyrinth** which acts as a statocyst; the labyrinth contains neuromast-type cells, perhaps derived during evolution from the neuromast cells of the lateral line. The labyrinth comprises three **semicircular canals**, set mutually at right angles to each other and filled with endolymph fluid. The canals connect to a **utriculus** and this connects with a **sacculus**, which are both small chambers. Plaques of hair cells in the utriculus and sacculus are covered with calcareous **otoliths**; hair cells in the canals form a crista, also with otoliths and covered with a gelatinous cupola, in the expanded ampulla at the end of the canals. **Gravity and acceleration** are detected by the otolith/hair cell system (as the animal moves forward, the inertia of the heavy otoliths retards them so that they push back against the hair cells; otoliths in the semicircular canals provide information about **position** and inform the CNS about **balance**.

Hearing and sound

Waves of alternating pressure in air (or water) are sensed as sound. Frequency determines pitch; pressure (proportional to amplitude) determines loudness. In insects, chitinous 'hairs' on the antennae form **sound-sensitive sensilla** (e.g. to

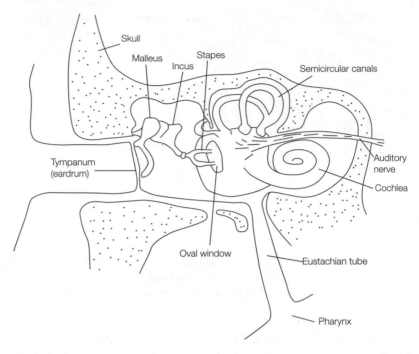

Fig. 4.   The structure of the human ear.

hear the stridulations of a cricket made when the insects rubs rough patches on the wings together). Some moths have **tympanic ears** with a drum of skin stretched across an air-filled chamber.

Fishes use their **lateral line systems** to detect low-frequency displacement waves in water: this is only efficient for sources nearby. Higher frequency pressure waves pass through the fish easily – thus the fish is transparent to sound. A large otolith in the sacculus may respond differently to these waves, and the **swim-bladder acts as a hydrophone**: pressure waves vibrate the bladder gases whose vibrations are relayed to the inner ear by a chain of Weberian ossicles.

Some terrestrial vertebrates (e.g. urodele amphibians, some lizards) rely on the conduction of low-frequency ground and air vibrations through bone. Anuran amphibians (e.g. toads), reptiles, birds and mammals use a tympanic ear.

The mammalian ear

In mammals, sound waves are collected by an outer ear or **pinna** which may be moveable (e.g. as in the dog). Waves travel down the **ear canal** (external auditory meatus) (*Fig. 4*) to the **tympanum** (eardrum) which is set into vibration. Vibrations are transferred by three **auditory ossicles**, the malleus, incus and stapes (hammer, anvil and stirrup), which act as a lever system in the middle ear and magnify the vibrations. (In reptiles the malleus and incus are involved in jaw articulation: in mammals they are incorporated into the ear.) The stapes pushes against the membrane covering the **oval window** leading to the cochlea of the inner ear. Since the area of the oval window is less than that of the tympanum, the vibrations are magnified further.

The **cochlea** (*Fig. 5*) has three fluid-filled canals separated by membranes: the upper and lower canals connect at the far end of the cochlea. As the oval window membrane vibrates, vibrations are transmitted to the cochlear fluid, setting up pressure waves. The waves travel to the far end of the cochlea and back to the round window of the lower canal whose membrane moves to complement that of the oval window, compensating for fluid movement.

The central canal of the cochlea has an **organ of Corti** resting on its base: this has sensory hair cells. Movement of fluid waves along the outer surface of the central canal causes vibrations in the basal membrane and so vibrations in hair cells. Vibration of a hair cell opens ion gates in its membrane, causing depolarization and release of neurotransmitters which trigger sensory neurons. Nerve impulses are interpreted as sounds within the auditory centers of the brain. The basal membrane is narrower and less elastic nearer the oval window and is sensitive to high-frequency vibrations, whereas the far end is wider and is sensitive to low-frequency vibrations. The human ear can detect sounds from about 16 Hz to 20 kHz (middle C = 256 Hz); during aging there is a loss of hearing at higher frequencies. Many animals can hear sounds in the ultrasonic range. Differential hearing in the two ears on each side of the head results in a perception of the direction of sound (stereophonic hearing).

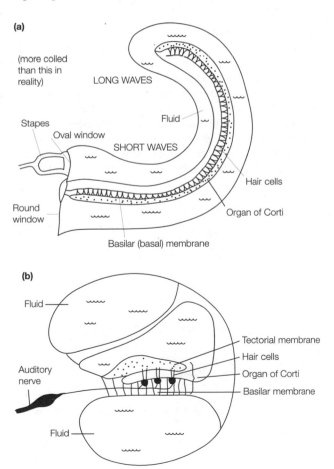

Fig. 5.   (a) The cochlea in the ear. (b) Cross-section of the cochlea in the ear.

Proprioception   Proprioceptors monitor the degree of stretch, contraction and tension in muscles and tendons. Golgi tendon organs (neuron endings) detect muscle tension. Many skeletal muscles contain spindles, receptors consisting of specialized (intrafusal) muscle fibers within a capsule. These are innervated by sensory and motor neurons: sensory messages about length changes and contraction are sent to the CNS, and appropriate motor responses may be initiated.

Photoreception   Light alters many organic molecules, and this forms the basis of photoreception.
and vision   Most animals possess photoreceptive cells containing a carotenoid pigment, **rhodopsin** (visual purple) or a variant. This absorbs light at wavelengths between about 400 and 800 nm. The pigments may be located in microvilli or (in vertebrates and echinoderms) in the membrane of the cilium of a uniciliate cell. In primitive forms, photoreceptive cells may be located over the body surface: passing shadows will elicit responses in the animal (e.g. earthworms). Usually photoreceptive cells are concentrated in an ocellus or eye. There is a sequence of complexity:

● Presence or absence of light
● Source and directional movement of light
● Focused images with or without color

Light-orienting eyes
Eyespots in euglenoid protoctistans have photoreceptive areas shielded by pigment on one side, allowing reception of signals interpreted by movement towards light. Flatworms have inverted pigment cup eyes (ocelli) (*Fig. 6*): light passes through a group of sensory neurons to reach a cup of pigment cells adjacent to the distal ends of the neurons. The worm tends to move so that the pigment cup is shielded from the light. Some polychaete worms and many molluscs and arthropods have several pairs of everted retinal cup eyes. The photoreceptive ends of sensory neurons are directed *away* from the light and form a light-sensitive retina. The cup is filled with a secreted 'soft lens'.

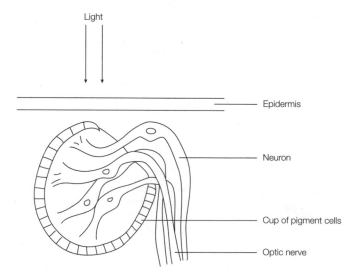

Fig. 6.   Diagram of a flatworm inverted eye.

Image-forming eyes

Image-forming eyes need a lens to focus divergent light rays. The more convex (or spherical) the lens, the greater the refracting (focusing) power. Aquatic animals (where the cornea has the same density as water) tend to have nearly spherical lenses since nearly all refraction occurs within them. Terrestrial animals (where the cornea is denser than air) have thinner lenses because most refraction occurs in the cornea. Image-forming eyes also need more photo-receptor cells.

Large image-forming eyes are found in cephalopods (e.g. *Octopus* sp.). The retina has 7×10^5 receptors per mm^2. Unlike vertebrate eyes (which cephalopod eyes superficially resemble), the octopus eye is everted, the photoreceptive cells being on the surface of the retina, facing the lens.

The vertebrate eye

In most species, the eyeball is more or less spherical (*Fig. 7*). Light from an object being viewed passes through the transparent **cornea** [which is kept moist and clean in terrestrial vertebrates by **lachrymal (tear) secretions**]. Rays are refracted by the cornea in terrestrial species; they are (further) focused by the **lens**. Lens muscles can change the shape (and so the focal length) of the lens to accommodate light rays with differing divergences from objects at differing distances. During aging, the lens muscles are less efficient at chang-ing lens shape, leading to **presbyopia** (far-sight). Defects in the shape of the eye also pose focusing problems as in **long-sightedness** (hypermetropia) and **short-sightedness** (myopia) (*Fig. 8*). **Astigmatism** is the result of a non-spherical eye.

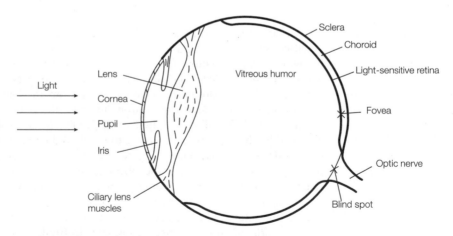

Fig. 7.   The structure of the vertebrate eye.

Stereoscopic vision depends on viewing the same visual field with two eyes simultaneously: each eye sees a slightly different view whose disparity is computed by the brain to give a perception of depth.

The amount of light is regulated by the perforated **iris** diaphragm. The size of the aperture (**pupil**) is regulated by muscles controlled by the autonomic nervous system. The eye is bounded by a tough **sclera** enclosing a vascular-ized **choroid** which is in turn lined by the **retina**. The cavities are filled with **humors**: the aqueous in front of the lens and the vitreous behind.

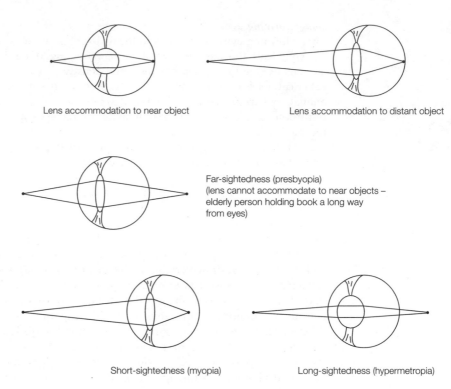

Lens accommodation to near object

Lens accommodation to distant object

Far-sightedness (presbyopia)
(lens cannot accommodate to near objects –
elderly person holding book a long way
from eyes)

Short-sightedness (myopia)

Long-sightedness (hypermetropia)

Fig. 8.   Accommodation in the eye and focusing problems.

An inverted image is focused on the retina. The retina has two types of light-sensitive cells: **rods**, for black-and-white vision, more sensitive to dim light, as in night vision; and **cones**, for color vision, with separate cones for blue, green and red primary colors. The absence of a cone gives a particular type of color blindness. The photoreceptive cells lie away from the lens so that light must pass through several layers of neurons to reach the photoreceptive cells. Only about 10% of light falling on the cornea reaches the rods and cones, most is absorbed by the black pigment lining the back of the retina. Photoreceptive cells communicate, via interneurons, to sensory neurons in the optic nerve. Where the neurons leave the eye is the **blind spot** with no rods or cones, compensated for by the other eye in humans; most photoreceptive cells are at the **fovea** (yellow spot) where vision is focused. Rods tend to be concentrated around the edge of the retina, whereas cones are situated more centrally.

In rods, **rhodopsin** comprises a pigment (**retinol**, derived from vitamin A) coupled to a protein (**opsin**). In the unstimulated cell, *cis*-retinol exists; light isomerizes it to *trans*-retinol which cannot bind to opsin. The broken rhodopsin affects the membrane permeability to sodium ions and so activates the photoreceptive cell. Following stimulation, *cis*-retinol is re-formed in the dark. Cone cells contain **iodopsin**.

Compound eyes
Insects and many crustaceans are characterized by compound eyes with many **ommatidia** units. Each ommatidium (*Fig. 9*) has a corneal lens where most refraction occurs. Deep cells in the cornea form a long crystalline cone which

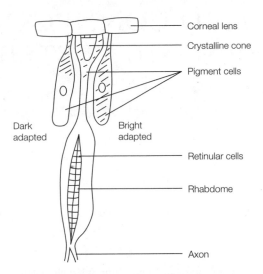

Fig. 9.   An insect ommatidium in the compound eye.

concentrates light onto elongated retinular cells (modified neurons). Pigment cells surround the cone and the retinular cells to prevent light from straying from one ommatidium to the next: the pigment may be retracted in dim light. The image formed by a compound eye is a rather coarse mosaic of spots of light detected by each ommatidium; resolution is thus poorer than in an image-forming eye, but the eye is very sensitive to perception of movement. Bees are capable of color vision, having retinular cells with pigments which absorb in the blue, green and ultraviolet parts of the spectrum.

Thermoreception   Temperature is often sensed by free nerve endings, although some cutaneous thermoreceptors in mammalian skin may be encapsulated. Some snakes (e.g. pit viper) have pits on their heads containing many thermoreceptors: these are sensitive to infra-red radiation, so that a snake can detect prey some distance away (e.g. a rattlesnake can detect, within 0.5 second, a rat 40 cm away with a body temperature only 10ºC above ambient temperature). Thermal intero-receptors are found in the hypothalamus of homoiotherms.

C26 BONE AND CONNECTIVE TISSUE IN THE VERTEBRATE SKELETON

Key Notes

Connective tissue

Connective tissue has few cells, a rich blood supply and extensive intercellular minerals and/or protein. It is divisible into connective tissue proper, cartilage and bone.

Connective tissue proper

This comprises fibroblasts in a fairly fluid intercellular matrix. It includes loose areolar tissue (general connective tissue), adipose tissue (with fat), dense connective tissue with collagen (muscle fasciae, ligaments), elastic connective tissue (arterial walls) and reticular connective tissue (spleen and liver framework).

Cartilage

Cartilage is a gel matrix with chondrocytes and fibers. Hyaline cartilage is glassy and supportive, and is found at the ends of bones. Fibrocartilage with collagen forms intervertebral disks. Elastic cartilage is found in elastic fibers in the larynx.

Bone

Bone is the main skeletal component. Bone functions include support, protection, leverage, mineral storage, storage of red blood cell synthetic apparatus and storage of energy (as marrow fat).

Bone histology

Bone is a living tissue with osteoprogenitor cells, osteoblasts, osteocytes and osteoclasts. It is formed from intercellular mineral (mainly hydrated calcium phosphate) and protein (mainly collagen).

Bone morphology

Long bones (e.g. the femur) have a shaft and epiphyses (tipped with hyaline cartilage). They are covered with periosteum. In land vertebrates the central cavity is marrow.

Compact bone

Compact bone (e.g. in the shaft) has osteons with longitudinal Haversian canals surrounded by concentric lamellae. Located between the lamellae are lacunae containing osteocytes with radiating canaliculi.

Spongy bone

Spongy (cancellous) bone (e.g. near epiphyses of long bones) has a lattice of thin plates, or trabeculae, containing lacunae with osteocytes. Trabeculae are parallel to axial forces on bone.

Ossification

Ossification starts when the mesencymal cells coalesce to form osteoprogenitor cells. Intramembranous ossification occurs within fibrous membranes (e.g. fetal skull), when osteoblasts accumulate on the matrix and deposit calcium salts. Endochondrial ossification occurs within a cartilage

	'model' of bone, first in the shaft, then in the epiphyses, leaving epiphyseal plates.
Bone growth	Elongation of bone occurs when new cartilage cells are generated on the epiphyseal side of the plate, and older cells are replaced by bone. Growth in diameter occurs when osteoblasts lay down bone on the outside; osteoclasts erode bone from the inside.
Modeling	Bone modeling is a delicate homeostatic balance between bone material deposition and resorption.
Joints	Joints are categorized into fibrous, cartilaginous and synovial joints. Synovial cavities are surrounded by a sleeve with outer, fibrous connective tissue elements which are extended to form ligaments. The capsule is filled with fluid.
Movements of joints	The movements of joints are divisible into gliding, hinge, pivot, ellipsoidal, saddle and universal.

Related topics

Skeletons (B3)
Muscles (C27)
Locomotion: swimming (C30)

Locomotion: terrestrial locomotion (C31)
Locomotion: flight (C32)
Buoyancy (C33)

Connective tissue Connective tissue is abundant in the body: it is characterized by few cells, extensive intercellular mineral and/or protein, and a rich blood supply. It supports, protects and binds other cells together. Connective tissue comprises 'connective tissue proper', cartilage, bone and blood (see Topic C9 for blood).

Connective tissue proper 'Connective tissue proper' typically has elongated fibroblast cells and a more or less fluid intercellular material. There are five types:

(1) **loose areolar tissue** is found around organs, in mucous membranes and under the skin; hyaluronic acid is in the intercellular matrix with some fibers and cells;

(2) **adipose tissue** is areolar tissue with cells specialized for fat storage: it can grow throughout life;

(3) **dense connective tissue** has collagen fibers: it is found in tendons, ligaments, muscle fasciae, etc.;

(4) **elastic connective tissue** has branching elastic fibers: it is found in arterial walls, trachea, vocal cords, etc.;

(5) **reticular connective tissue** has a network of fibers: it forms the loose frame of the liver, spleen, lymph nodes, etc.

Cartilage Cartilage has a gel matrix containing chondrocytes and collagen and elastic fibers. It grows from within and without.

● **Hyaline cartilage** has a glassy consistency; it is found at the ends of bones, in the nose and the respiratory tract, and is supportive.

- **Fibrocartilage** is very strong. Strengthened with collagen, it connects bones (e.g. in the pelvis) and forms intervertebral discs (the relic of the notochord in most vertebrates).
- **Elastic cartilage** maintains the shapes of the external ear (pinna) and the larynx.

Bone

Bone is the main component of the skeletal system. Functions of bone (with cartilage) include:

- support (framework of body);
- protection (e.g. skull);
- leverage (limbs);
- mineral storage (especially calcium and phosphorus);
- storage of blood cell synthetic apparatus (red bone marrow: land vertebrates only);
- storage of energy (fat in yellow bone marrow: land vertebrates only).

Bone histology

Bone comprises many, widely separated cells surrounded by the intercellular matrix of bone tissue. Four principal bone cell types are found.

(1) **Osteoprogenitor** cells are stem cells which give rise to osteoblasts: they are found in the periosteum surrounding bones and in the endosteum membrane lining the inner bone cavity, also in canals in bone carrying blood vessels.
(2) **Osteoblasts** (on bone surfaces) secrete bone mineral and collagen.
(3) **Osteocytes** are osteoblasts within the bone matrix which they have built up around themselves; they maintain the functioning of bone.
(4) **Osteoclasts** (cells in the monocyte/macrophage series) resorb excess bone material.

The bone matrix has collagen fibers and crystalline mineral (mainly hydroxyapatite, a hydrated calcium phosphate with some carbonate, fluoride, sulfate and magnesium salts). The protein collagen makes up about 33% of bone: the minerals crystallize around the protein framework. During aging, the proportion of protein in bone decreases, making it more brittle. Note that bone is a *living* tissue.

Bone morphology

A typical long bone (*Fig. 1*) (e.g. femur) comprises a shaft (diaphysis) and ends (epiphyses) connected by metaphyses. The cylindrical structure of a long bone concentrates most material around the periphery (as seen in cross-section), thus giving maximum resistance to bending moments. A thin layer of hyaline cartilage covers the epiphyses and reduces joint friction. The bone is covered by a periosteum: this has an outer fibrous layer (with nerves and blood vessels) and an inner osteogenic layer with bone-forming cells. In the middle (in higher, land vertebrates) is the marrow cavity, separated from the bone material by the endosteum.

Compact bone

Compact bone has few cavities within it; it covers spongy bone (see below) and is particularly thick in the shaft. It has a concentric ring structure (*Fig. 2*). Vessels and nerves from the periosteum penetrate the bone via perforating canals and connect with nerves and vessels of the marrow cavity and the central **Haversian canals** which run longitudinally through the bone. Around the Haversian canals

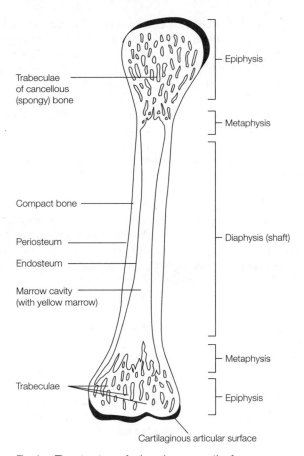

Fig. 1.   The structure of a long bone, e.g. the femur.

are concentric **lamellae**, rings of hard 'bone'. Between the lamellae are small spaces, **lacunae**, containing osteocytes.

Radiating from the lacunae are **canaliculi** which contain processes from the osteocytes and extracellular fluid; the canaliculi interconnect with each other and the Haversian canals. The Haversian canal plus the surrounding lamellae, lacunae, osteocytes and canaliculi form an **osteon**.

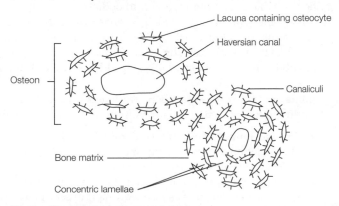

Fig. 2.   The structure of a compact bone.

Spongy bone

Spongy (cancellous) bone lacks true osteons; it comprises an irregular lattice of thin plates (**trabeculae**). Within the trabeculae are lacunae with osteocytes. Spongy bone is rich in blood vessels and marrow lies within the network of trabeculae. The trabeculae are aligned parallel to the main forces of compression and tension in the bone.

Ossification

Bone forms by **osteogenesis (ossification)**, starting when mesenchymal (see Topic D2) cells in the embryo transform into osteoprogenitor cells which differentiate into osteoblasts and osteoclasts.

Intramembranous ossification occurs within fibrous membranes of the embryo, fetus and child (e.g. skull bones). Clusters of osteoblasts accumulate on a fibrous matrix and deposit calcium salts.

Endochondrial ossification occurs within a cartilage model, the primary ossification center being the shaft. Breakdown of cartilage results in cavitation: cavities merge to form a marrow cavity. Osteoblasts deposit mineral and protein, eventually replacing cartilage at the epiphyses where only the cartilaginous articular end-sheets remain together with the **epiphyseal plates** between the epiphyses and the shaft: the latter are the sites for bone growth.

Bone growth

Bones **elongate** by **appositional growth** at the epiphyseal plates under hormonal control (e.g. growth hormone). New cartilage cells are generated on the epiphyseal side of the plate, and the older cartilage cells are destroyed and replaced by bone on the shaft side of the plate (thus the plate has a constant thickness but the length of the shaft increases).

Growth in diameter occurs when osteoblasts from the periosteum add new bone to the outer surface of the bone while osteoclasts erode bone material inside the shaft and so enlarge the marrow cavity.

Modeling

There is a homeostatic balance between formation and resorption of bone material, called **remodeling**. Calcium, phosphorus and vitamins C and D are essential. Overall bone growth is controlled by growth hormone from the anterior pituitary; calcitonin from the thyroid inhibits osteoclast activity; parathyroid hormone (parathormone, PTH) increases osteoclast activity; sex steroids promote a burst of bone growth during puberty followed by the destruction of the cartilage cells of the epiphyseal plates and epiphyseal fusion, with a cessation of bone growth in early adulthood.

Bone mechanically stressed in exercise can increase its strength due to the generation of piezoelectric currents (from mineral crystals) which stimulate osteoblasts. In aging, calcium is withdrawn from bones with a tendency towards osteoporosis; less collagen makes aged bones more brittle.

Joints

Joints are points of contact between bones (or cartilage and bone) allowing for varying degrees of movement, from none to freely movable.

- **Fibrous joints**: bones are held together with fibrous connective tissue. A fibrous joint permits minimal movement, for example sutures between skull bones, articulation of teeth in jaws.
- **Cartilaginous joints**: bones are held tightly together by cartilage and permit minimal movement. Examples are epiphyseal plates within bones and the pubic symphysis between the two halves of the pelvis.

● **Synovial joints**: these allow movement (*Fig. 3*). There is a **synovial cavity** between bones; hyaline cartilage covers the ends of the articulating bones. Joints are surrounded by a sleeve **capsule** with an outer, fibrous dense connective tissue layer attached to the periosteum. If the fibers are arranged in parallel bundles they are called **ligaments**. The inner capsule layer is the synovial membrane of looser connective tissue: this secretes a lubricating **synovial fluid**. Pads of fibrocartilage (articular disks on the articular surfaces of the bones) may be present within the synovial capsule. (Similar capsules, bursae, may be found in the body at sites of friction, e.g. where skin rubs over bone.)

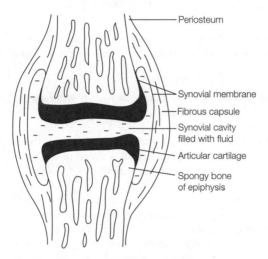

Fig. 3.   The structure of a synovial joint, e.g. the knee.

Movements of joints

Joints may be further categorized into:

● **gliding joints** (usually flat, e.g. ribs over vertebrae);
● **hinge joints** (angular movement, e.g. elbow);
● **pivot joints** (rotation, e.g. atlas vertebra allowing head-shaking);
● **ellipsoidal joints** (side-to-side or back-and-forth, e.g. wrist);
● **saddle joints** (freer version of ellipsoidal joint, e.g. base of thumb);
● **universal joints** (ball-and-socket, e.g. hip).

C27 MUSCLES

Key Notes

Muscle cells	Muscle cells are specialized for contraction, and can be found throughout the Animal Kingdom. Muscle cells are often organized and differentiated into muscular tissue.
Properties of muscle	Muscles generate force by contraction. Muscle cells are excitable, contractile, extensible and elastic. Muscles are found usually in antagonistic pairs around skeletal elements.
Functions of muscle	Functions include motion, maintenance of posture and heat generation (especially in homoiotherms).
Muscles types in mammals	Striped (striated, skeletal, voluntary) muscle moves the skeleton. Cardiac muscle (involuntary striped muscle) is found in the heart. Smooth muscle (involuntary) surrounds internal organs.
Striped muscles	These are surrounded by connective tissue fasciae; bundles of muscle cells (fasciculi) and fibers (cells) are themselves also surrounded by connective tissue. Connective tissue extends to form tendons. Muscles are innervated and richly vascularized. Fibers (cells), containing parallel myofibrils, are bounded by the sarcolemma (plasma membrane). Each myofibril has a bundle of myofilaments divided longitudinally into sarcomeres by Z lines. Sarcomeres comprise heavy (myosin) and light (actin) proteins; in cross-section these are lined up in a hexagonal array. The optical appearance of overlapping regions gives striations with I, H and A bands.
Sliding filament model	Myofilaments slide past each other as the myosin head attaches to the actin filament. This changes the conformation and pulls the actin filament along in a ratchet fashion.
Energy for contraction	ATP provides immediate energy; phosphocreatine can be used as a further short-term energy store to regenerate the ATP pool. Longer-term energy needs are provided by glycolysis, aerobic oxidation of pyruvate or anaerobic metabolism of glucose to lactate.
Regulation of contraction	Tropomyosin blocks myosin-binding sites on actin. When troponin binds calcium it can displace tropomyosin to allow actin–myosin interaction. The action potential in the sarcolemma is transmitted via T tubules (indented sarcolemma) throughout the fiber. The sarcoplasmic reticulum sequesters calcium. The action potential stimulates calcium release from the sarcoplasmic reticulum, and calcium binds to troponin. After the potential has decayed, calcium is actively pumped back into the sarcoplasmic reticulum.

Cardiac muscle	This resembles striped muscle but the fibers are shorter and have branched ends and communicate via gap junctions to intercalated disks. Spontaneous, involuntary contractions are seen.
Smooth muscle	Actin and myosin myofilaments are less regularly arranged; calcium is stored in extracellular fluid. Involuntary contractions occur; muscles are arranged in sheets around hollow organs with longitudinal muscles outside and circular muscles inside. Multi-unit fibers have each fiber with a motor neuron end-plate; visceral smooth muscle fibers have few fibers with motor neuron end plates: transmission occurs via gap junctions.
Muscle contraction physiology	The neuron plus fibers is a motor unit. Nerve impulses cause all fibers in a motor unit to contract simultaneously. Contraction against a constant load is isotonic, whereas contraction against an immovable object is isometric.
Phasic and tonic muscles	Phasic (twitch) muscles have fibers each with one motor nerve end-plate. Single stimulation of threshold intensity leads to a sarcolemma action potential and an all-or-nothing twitch (gradation is effected by temporal summation, possibily leading to a state of tetanus, or, in the whole muscle, by spatial summation). Slow phasic fibers maintain posture: these are rich in myoglobin and mitochondria, and are important in aerobic metabolism. Fast phasic (glycolytic) fibers are for quick bursts of movement: these lack myoglobin, have fewer mitochondria and are important in anaerobic metabolism. Tonic muscles have many motor neuron end-plates per fiber. There is no sarcolemma action potential and tension is proportional to sarcolemma depolarization.
Arthropod muscles	There are few neurons per muscle and one neuron can innervate many muscles. Fibers may have multiple motor neuron end-plates from up to five neurons; at least one is inhibitory.
Muscle work and power	Force relates to the number of actin–myosin cross-bridges, and is proportional to the cross-sectional area. Work is the product of force and distance. Longitudinal (fusiform) and pennate fiber patterns balance the distance of contraction against the number of fibers. Power is the rate of doing work. Small muscles which contract quickly can be very powerful per unit weight of muscle. Maximum power develops at intermediate velocities.
Evolutionary aspects	Somatic muscles relate to the body wall and appendages, visceral muscles relate to the gut and are usually smooth. Fish somatic muscles comprise segmented myomeres and abductor and adductor fin muscles. These are extensively modified in tetrapods.

Related topics

Muscle cells

Muscle cells are specialized for contraction: every muscular function, from running and jumping, to ventilating the gills or lungs, propelling blood through the vascular system or food through the gut, or expelling an egg from the oviduct, is carried out by muscle cells working in concert.

Many protoctistans have contractile stalks (e.g. the ciliophoran *Stentor* sp.), and contractile cells occur in sponges associated with the openings (ostia) through which water is discharged into the spongocoel central cavity.

Contractile muscle cells are found in the column walls of Cnidaria such as sea-anenomes. In higher animals, single muscle cells persist such as the myo-epithelial cells associated with mammalian sweat glands. Most contractile cells in Metazoa are organized and differentiated into muscular tissue.

Properties of muscle

Muscles generate force. In humans they comprise 40–50% of the total body weight. Muscle is:

● **excitable**: it receives and responds to stimuli;
● **contractile**: it shortens and thickens to do work;
● **extensible**: it can be stretched passively when relaxed;
● **elastic**: it returns to its original shape after contraction and extension.

Note that muscles only do work when they contract: contraction is the active process.

Most muscles associated with the skeleton are arranged in opposing, antagonistic pairs: when one contracts the other is passively stretched (e.g. the biceps and the triceps bending and straightening the human elbow). In bivalve molluscs, the closing adductor muscles work against the elastic spring ligament in the hinge of the shell.

Functions of muscle

There are three major functions for muscle:

(1) **motion**, including locomotion;
(2) maintenance of **posture;**
(3) **heat generation** (particularly in homoiotherms): approximately 85% of human body heat is generated by muscle contractions; brief involuntary muscle contractions (shivering) generate heat when the animal is cold.

Muscle types in mammals

There are three types of muscle tissue.

(1) **Striped (striated) muscle** moves the skeleton; often called voluntary (because its movement is usually voluntary) or skeletal muscle.
(2) **Cardiac muscle** is a special type of striped muscle in the walls of the heart; contraction is involuntary.
(3) **Smooth muscle** surrounds the walls of internal organs such as the gut, bladder, blood vessels and the uterus; contraction is involuntary.

Striped muscle

The disposition of striped (skeletal or striated) muscles and their names is best studied using an illustrated guide [e.g. Clancy and McVicar (1995); see Further Reading].

Striped muscles are surrounded by superficial **fasciae** composed of connective and adipose tissues: they provide a route for nerves and blood vessels. **Deep fascia** is connective tissue which holds muscles together and separates them into functioning units. The entire muscle is also wrapped in fibrous connective tissue: the **epimysium**.

Bundles of muscle cells (**fasciculi**) are covered by **perimysium**. **Endomysium** penetrates into the fascicles and surrounds and separates the muscle cells. The epimysium, perimysium and endomysium contribute collagenous fibers to tendons which attach muscle to the periosteum of bone. Tendons can be flat or broadly cylindrical.

Muscles are richly vascularized to deliver oxygen and glucose and to remove waste products; they are also extensively innervated. The muscle fibers (or cells, bounded by a plasma membrane or **sarcolemma**) contain contractile organelles called **myofibrils** (*Fig. 1*); each has a bundle of **myofilaments** divided lengthways into repeating structural units called **sarcomeres**. The sarcomeres are divided into bands of different densities, giving the striped appearance characteristic of this muscle type.

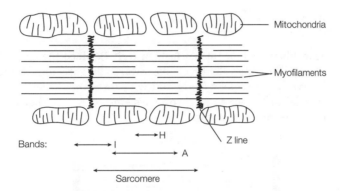

Fig. 1.   The structure of skeletal (striped) muscle: a myofibril.

There are two kinds of myofilaments:

- heavy **myosin** filaments;
- light **actin** filaments.

X-ray crystallography shows that, in cross-section, the filaments line up in hexagonal array. Actin filaments are held by their attachment to the **Z lines** (made up of the protein α-actinin). **I bands** are regions containing actin filaments alone; the light **H band** in the middle of each sarcomere is where only the heavy myosin filaments are present; the myosin and actin filaments overlap in the dark **A bands**.

Sliding filament model

When a striped muscle contracts, the filaments (which do not change length) slide past each other. In each sarcomere, the many globular myosin heads which project laterally along each end of the heavy myosin filament attach to the actin filament and change conformation. The myosin pulls at the actin filaments adjacent to it. The myosin heads have been energetically charged, adopting a conformation in which they can bind to actin. This binding elicits the conformational change that provides the force for filament sliding and exposes an ATP-binding site. ATP binding causes an allosteric (shape) change that promotes detachment of the head from actin. Dephosphorylation of ATP provides the energy to re-establish the actin binding; thus the process is repeated many times (each using one ATP molecule) and the myosin pulls along the

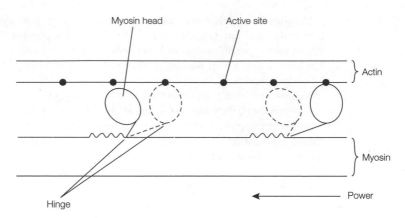

Fig. 2.   The action of the actin–myosin ratchet.

actin filament in a ratchet fashion (*Fig. 2*). Since the ends of each myosin filament pull in opposite directions, towards the sarcomere center, the myosin pulls
the two actin regions closer and with them the Z lines (*Fig. 3*): thus the whole
muscle contracts. (Actin–myosin systems are found elsewhere: e.g. pulling chromosomes apart at mitosis.)

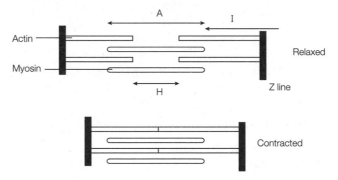

Fig. 3.   The relationship of actin and myosin filaments in relaxed and contracted striped
muscle.

**Energy for
contraction**

The **energy** for contraction is provided by **ATP**: the myosin heads have an
ATPase, the active site of which is exposed upon actin binding. The ATP synthesized and stored in muscle can only sustain a few seconds of vigorous exercise.
Striped muscles contain high levels of **phosphocreatine**: this breaks down to
release energy [some of which is used to make more ATP from adenosine
diphosphate (ADP) and the phosphate is released]:

$$\text{Phosphocreatine} + \text{ADP} \rightleftharpoons \text{Creatine} + \text{ATP}$$

Longer-term energy needs must be supplied by cellular respiration:

- breakdown of glycogen (via glucose) to pyruvate with the synthesis of ATP,
  used as above: this is **glycolysis** and does not require oxygen;
- oxidation of pyruvate, in the presence of oxygen, to yield carbon dioxide,
  water and large amounts of ATP (38 molecules of ATP following the total
  oxidation of one glucose molecule: this is **aerobic respiration**).

In the absence of oxygen, the pyruvate is normally converted to lactate: this is **anaerobic metabolism**. In heavy exercise, the lactate spills over into the blood. In humans, the glycogen/lactic acid system can provide energy for about 40 seconds of maximum muscle activity. The lactate constitutes an 'oxygen debt' which must later be paid off by oxidizing the lactate to carbon dioxide and water.

Aerobic respiration can continue more or less indefinitely, provided oxygen and glucose (or another fuel such as fats) are available

Regulation of contraction

Regulation of contraction is effected primarily by the proteins **tropomyosin** and **troponin**. Tropomyosin blocks the actin sites to which myosin will attach. Troponin has sites which bind **calcium**: this changes the troponin shape, so allowing the troponin to displace ('push aside') the tropomyosin in order that actin and myosin can interact.

Neuromuscular transmitter release (acetylcholine from the endings of motor neurons innervating striped muscle in mammals) triggers an action potential in the plasma membrane (sarcolemma) of the muscle fiber (cell). The **T (transverse) tubule system** is continuous with the cell membrane (sarcolemma), and runs through the muscle fibers close to the Z lines. Thus action potentials can be rapidly transmitted into the fibers.

The **sarcoplasmic reticulum** is a second system which encloses the myofibrils. It contains calcium ions, and by sequestering them it keeps the concentration around the myofibrils low. The reticulum expands near the Z lines of each sarcomere where the reticulum is in intimate contact with the T tubule system (although the two systems are not physically continuous).

The action potential is transmitted along the membranes of the T tubules (propagation is similar to that along the plasma membrane of a neuron) and stimulates calcium release from the sarcoplasmic reticulum, so that the intracellular level of calcium increases 100-fold; the calcium initiates contraction by binding to troponin (see above).

Immediately after the action potential has activated the system, the calcium is pumped back into the sarcoplasmic reticulum (an ATP-dependent process); acetylcholinase (or acetylcholinesterase) breaks down the neuromuscular transmitter acetylcholine. Each contraction lasts for a few tenths of a second, until the calcium pumps have reduced the intracellular calcium pool to a point where the contractile apparatus is no longer operational; repeated action potentials cause long-term contraction or muscle tonus or tetanus.

Cardiac muscle

Cardiac muscle resembles striped muscle in that it has similar assemblies of actin and myosin and has a striated appearance. The fibers are shorter and have branched ends. Contraction is spontaneous and involuntary. Gap junctions in the **intercalated disks** that join the branched cells longitudinally transmit action potentials so that the whole of the muscle contracts synchronously.

Smooth muscle

Smooth, involuntary muscle is also made up of actin and myosin, but the molecules are arranged much less regularly so there are no striations (thus 'smooth'). Actin–myosin interactions are similar to those in striped muscle, although the actin filaments are attached to dense bodies rather than to Z lines. Calcium is not stored in a sarcoplasmic reticulum, but in the extracellular fluid. Contractions tend to be slower and more prolonged. In hollow organs such as the gut, the muscle fibers are arranged in sheets with a longitudinal orientation outside and a circular orientation inside.

In vertebrates, smooth muscle fibers are divisible into two major types.

(1) **Multi-unit smooth muscle** fibers (e.g. in the iris of the eye): here nerve impulses reach each fiber from a motor nerve end-plate.
(2) **Visceral smooth muscle** fibers (e.g. gut and uterus walls): here only a few fibers have motor nerve end-plates, the action potential flowing to adjacent fibers through gap junctions. Such muscle fibers will contract spontaneously when stretched.

Muscle contraction physiology

For most vertebrate striped muscles, a motor neuron branches so that there are several motor end-plates each on one of several muscle fibers (cells): the neuron plus its fibers is the **motor unit**. A nerve impulse causes all fibers in the unit to contract simultaneously. The more fibers per unit, the less the control which can be exercized (some eyeball muscles may have only two or three fibers per neuron while locomotory muscles may have 400–500 fibers per neuron).

Contraction of a fiber may be initiated by a single nerve impulse, or several impulses with a summating effect may be necessary. There is a latent period of a few milliseconds between the arrival of the impulse and actin–myosin interaction. Tension develops during the contraction period and then decreases during the longer relaxation period.

If a muscle works against a constant load so that the muscle shortens, this is **isotonic** contraction (e.g. lifting this book from the desk) or positive work; if the muscle pulls against an immovable object (e.g. pulling at a locked door), this is **isometric** contraction: internal shortening and tension develop as the elastic elements in the muscle are stretched but the muscle does not shorten.

Phasic and tonic muscles

Most striped muscles in vertebrates are **phasic (twitch) muscles**. Each fiber has one motor nerve end-plate. When the motor unit is stimulated by its neuron, a single stimulus with a threshold intensity leads to a **twitch** which is all-or-nothing because the action potential spreads rapidly throughout the whole fiber. Gradation in muscle tension is facilitated by:

● imposing a second twitch on the first before the latter's effects have decayed (not all the myosin heads are activated during the time available for the first twitch) this is **temporal summation**; or
● delivering impulses to reach the motor unit so rapidly that no relaxation can occur between action potentials, so eliciting a plateau of tension or state of **tetanus** which persists until the impulses ceases or the muscle fatigues; or
● progressively recruiting more motor units to contract: this is **spatial summation**.

Some muscle fibers (but not the same ones continuously) will always be contracted in a given muscle: this is called **tonus** and allows the muscle to be held firm.

Slow phasic fibers are used to maintain posture and in endurance activities: they are often rich in myoglobin and are dark red in color. Such fibers are rich in mitochondria. Their slow contraction allows complete, aerobic oxidation of fuels and they fatigue only slowly. **Fast phasic (glycolytic) fibers** are used for quick bursts of movement; they lack myoglobin and have fewer mitochondria; their metabolism is often anaerobic. The two fiber types may be mixed (as in most mammalian muscles) or separated (the red and white meat in chicken or herring).

Tonic muscles are like phasic muscles with slow fibers, except that they have many nerve end-plates per fiber and do not follow the all-or-nothing phenomenon seen in a phasic muscle fiber twitch. Action potentials do not spread far and gradation in contraction is effected by more frequent impulses. Bivalve (e.g. oyster) adductor muscles contain phasic and tonic fibers, the former snapping the valves of the shell closed while the tonic fibers keep it closed.

Arthropod muscles

Arthropod muscles (e.g. in crab claws) have very few neurons per muscle, and a given neuron may innervate several muscles. Each fiber may have multiple end-plates from each motor neuron, and up to five neurons may innervate one fiber: at least one neuron is inhibitory whereas the others facilitate varying degrees of slow or fast contraction. Combinations of motor stimulation can thus permit a gradation of action.

Muscle work and power

Muscle **force** reflects the number of actin–myosin cross-bridges formed, in turn dependent on the number of fibers within the muscle, roughly proportional to its cross-sectional area:

$$\text{Force} \propto \text{Cross-sectional area of the muscle}$$

Force of contraction per unit cross-sectional area is usually between 4 and 6 kg cm^{-2}. Muscles produce most force when contracting around their resting length.

Work is the product of force and the distance through which the force works: it is more or less constant per gram of muscle, but large muscles will, of course, generate more force. In some muscles, the fibers are arranged longitudinally (fusiform pattern) which permits contraction over a greater distance compared with muscles with the pennate or feathered pattern of fibers where the fibers are arranged to allow shorter fiber contraction, but the larger number of fibers generate much more force. Thus large muscles (e.g. the quadriceps of the human thigh), with a large cross-sectional area and therefore force, which can contract significant distances (e.g. about one-third the resting length), can do large amounts of work compared with small, short muscles.

Power is the rate of doing work. Very small muscles (e.g. those of the eyeball in a small rodent) are very powerful in that they contract very quickly: the power output per unit weight of muscle is higher than that of the equivalent, larger muscles of a larger mammal.

Muscles develop their maximum power at intermediate velocities: the actual velocity depends on whether the fibers are phasic or tonic, but efficiency is at a maximum at 20–30% of maximum velocity. Energy is lost because of inefficiencies in the energy interconversion processes: much will be lost as heat and to overcome internal friction in the muscles.

Evolutionary aspects

Somatic muscles in vertebrates are associated with the body wall and appendages (fins and limbs) and are usually striped; **visceral muscles** are associated with the gut and are usually smooth. Fish somatic muscles mainly comprise segmental **myomeres** which facilitate the undulation of swimming.

Paired fins are moved by a dorsal abductor muscle (pulling the fin up) and a ventral adductor muscle (pulling it down). On land, the segmental arrangement of muscles is largely lost. The fin abductor and adductor muscles become divided to form the limb muscles attached to the limb girdles, and additional muscles develop in the limbs themselves.

C28 NONMUSCULAR MOVEMENT

Key Notes

Cytoskeleton	Cells contain cytoskeletal elements with contractile actin and myosin filaments, and microtubules.
Amoeboid movement and pseudopodia	A fluid inner endoplasm with unpolymerized actin flows forward in a pseudopodium to the fountain zone where actin polymerizes and contributes to the outer ectoplasm.
Flagella and cilia	Tubulin microtubules (in a 9 + 2 arrangment) with dynein arms bring about bending of flagella and cilia, leading to propulsive reactions against the medium.
Related topics	The Protozoa (A1) Muscles (C27)

Cytoskeleton

Most cells contain cytoskeletal elements comprising contractile filaments of actin and myosin, together with microtubules. This permits shape changes in the cells and facilitates the evolution of specialized mechanisms for movements.

Amoeboid movement and pseudopodia

Characteristic of protoctistans such as *Amoeba* spp. but also found in cells as diverse as amoebocytes of sponges and macrophages in mammalian blood and tissues, amoeboid movement is associated with pseudopodial processes on the cell surface.

Pseudopodia ('false feet') can be used to engulf food particles or can attach to the substratum and pull the cell or organism along. Several psuedopodia can form on a cell surface simultaneously, but one becomes dominant. The pseudopodium has fluid **endoplasm** on the inside and stiffer **ectoplasm** as a sleeve outside. The deepest endoplasm moves 'forward' (i.e. towards the front of the pseudopodium which may not correspond to the front of the organism or cell), separated from the ectoplasm by a shear zone. At the 'front' of the pseudopodium it spreads in the fountain zone where it becomes more rigid and contributes to the advancing ectoplasm sleeve. At the rear end, ectoplasm is recruited into forward-moving endoplasm (*Fig. 1*).

In the endoplasm, actin is unpolymerized. Calcium, ATP and a binding protein facilitate the polymerization of actin in the fountain zone.

Flagella and cilia

Flagella and/or cilia are found in protoctistans and almost universally in animals. No motile cilia are found in nematodes and arthropods (although processes on arthropod sensory cells are modified cilia).

Flagella

Usually, undulatory waves beat from the base to the tip of the long, whip-like flagellum; reaction to the beat in small components of the flagellum can be

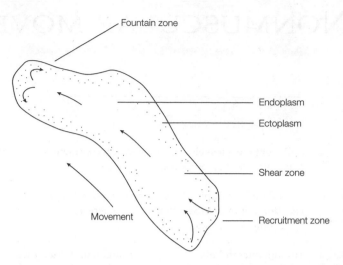

Fig. 1.   Pseudopodial movement.

resolved into lateral and propulsive components – the lateral components cancel out, giving net propulsive thrust (*Fig. 2*). Successive waves are in the same plane or at 90º to each other, giving a helical, propellor-like thrust. On some flagella, processes are found enabling the flagellum to pull rather than push, this also affects the direction of the wave which is now **direct** rather than **retrograde**.

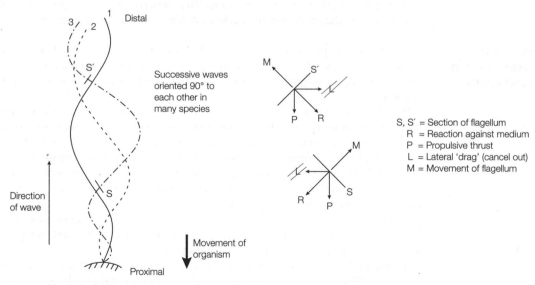

Fig. 2.   Flagellar movement.

Cilia

The beat of cilia is more planar; the action is oar-like. The cilium stiffens for the propulsive stroke but is more flexible for a 'feathering' recovery (*Fig. 3*).

Cilia and flagella have a common and universal structure in eukaryotes. In cross-section, nine circumferentially arranged double-microtubules (doublets)

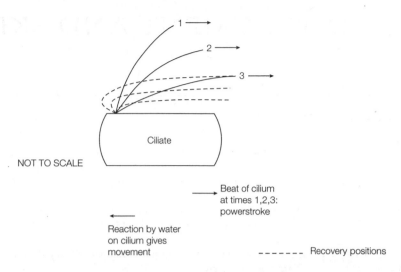

Fig. 3. *Ciliary movement.*

are seen together with two central, single microtubules (the **'9 + 2'** model) (*Fig. 4*). Radial spokes exist between the outer doublets and the core tubules. The tubule walls are made of the protein **tubulin**. **Dynein** arms form temporary links to the next doublet. ATP-powered formation and breakage of dynein links allows microtubules to slide past each other, one doublet ratcheting past the next. The radial spokes permit the transformation of the sliding action into bending.

At the bases of cilia and flagella are organizing **kinetosomes** (basal bodies). These are equivalent to the centrioles associated with mitotic spindles.

Where there are many cilia (e.g. in ciliophoran protoctistans such as *Paramecium* sp.), the ciliary beating is externally co-ordinated by hydrodynamic forces into a **metachronal wave**. Beat reversal (under internal control) is possible. Some metazoan animals (e.g. nemertine ribbonworms) use cilia to creep over the substratum: a layer of mucus may lubricate the surface. The mucus can itself by moved by cilia and used to trap food (e.g. in the sea-squirt pharynx) or to remove debris (e.g. in the mammalian trachea).

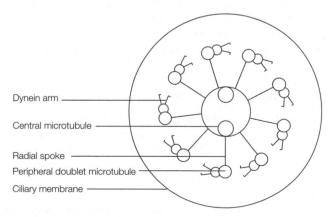

Fig. 4. *The 9+2 model of the cross-section of a cilium.*

C29 INTEGUMENT AND SKIN

Key Notes

Integument	The integument separates the internal and external environment: it protects and supports. It may be associated with controlled transfer of substances into and out of body, control of heat loss and gain, and sensory receptors.
Epithelium	Close-packed epithelial cells lie on a basement membrane; together with connective tissue they make up the integument.
Invertebrate integuments	This is a single layer of columnar epithelial cells (epidermis) which may be ciliated or secrete cuticle or shell.
Vertebrate integuments	The skin comprises the epidermis with a lower germinal layer, and a connective tissue dermis. Keratin protein may be present in the epidermis, forming scales, feathers, hair or horns. Fish epidermis lacks keratin – exposed bones in folds in the dermis form scales. The stratum corneum in terrestrial vertebrates forms an extensive waterproofing keratin layer which shows as reptile scales and bird feathers. In mammalian skin, keratin hairs emerge from epidermal follicles and there are extensive glands.
Related topics	Thermoregulation (C21) Bone and connective tissue in the vertebrate skeleton (C26)

Integument

Integuments separate animals' internal and external environments. They may consist of a single layer of cells or have many layers.

The functions of the integument include:

- protection against pathogenic invasion;
- protection against ionizing radiation (including ultraviolet rays);
- protection against desiccation or osmotic shock;
- support or maintenance of body turgor;
- protection against predation [including mechanical protection (e.g. porcupine), warning coloration (e.g. arrow-poison frog) and camouflage (e.g. leopard)];
- protection against abrasion.

Associated with the integument may be the transfer of gases, ions, water and wastes. Receptor cells to detect changes in the external environment are situated in or near the integument, and control of heat loss and gain, particularly in homoiotherms through feathers, hair, fur, blubber or exposed surfaces, are also integumentary features.

Epithelium

Epithelium and connective tissue make up the integument. Epithelial cells are closely placed and may be cuboidal, columnar or squamous (flattened): they are also found lining internal cavities such as the gut and blood vessels, and in glands (e.g. the pancreas). Epithelial cells rest on a basement membrane, a fibrous, extracellular sheet. Some epithelial cells exist in single-layer sheets and this is termed simple epithelium; stratified epithelium consists of two or more layers of cells (see Topic C27).

Invertebrate integuments

Invertebrate integuments usually comprise a single layer of columnar epithelial cells, termed an **epidermis**, resting on a basement membrane. Glands may secrete a tough cuticle which helps to maintain body turgor (e.g. in nematodes) or a shell (e.g. in crustaceans). Cilia may be present on exposed epidermal cells (e.g. in ribbonworms).

Vertebrate integuments

The vertebrate integument (skin) comprises, from the outside:

● epidermis (stratified, squamous epithelium);
● germinal layer (stratum germinatium): the lowest layer of the epidermis and comprising cuboidal cells actively dividing; the progeny of these divisions move to the surface, become squamous, and are shed;
● dermis (thick layer of dense connective tissue, rich in collagen fibers); the dermis may contain nerves, sensory receptors and blood vessels; bone may form in the dermis.

Epithelial cells may accumulate **keratin**, a protein which forms scales, feathers, horn, nails, claws and hair, the cells dying in the process. Fish epidermis is thin and has no keratin. It is often rich in mucus cells; mucus lowers friction and acts as an antimicrobial barrier. Overlapping folds are frequently found. Bony scales or spines develop within the fold dermis and as the epidermis erodes the scales are exposed. Modern amphibian skin lacks scales although keratin may be present.

Terrestrial vertebrates synthesize increased amounts of keratin which is important for waterproofing the body: keratin is particularly present in a horny **stratum corneum**. In reptiles, the stratum corneum forms **horny scales**. In birds these scales are modified to form **feathers**.

In mammals, **hairs** emerge from follicles of epidermal cells which indent the dermis. **Glands** are infrequent in reptile skin, but mammalian skin glands (developing from ingrowths of the epidermis into the dermis) include:

● eccrine sweat glands (produce sweat to cool the body);
● apocrine sweat glands (produce odoriferous sweat-like secretions from the axilla and pubis);
● sebaceous glands (secrete oil into hair follicles);
● mammary glands (secrete milk: may be modified sweat or sebaceous glands).

Color in the integument may be due to:

● vascularization in the dermis;
● iridescent granules (e.g. beetles);
● pigments in star-shaped chromatophores (e.g. frogs): movement of pigment such as black melanin, to and from the centers of the chromatophores, under nervous control, can effect color changes in the skin.

C30 LOCOMOTION: SWIMMING

Key Notes

Locomotory principles	Movement depends on an animal pushing against a fluid medium or the substratum to generate a reaction which is translated into locomotion.
Changes in body shape	Jet-propulsion is used by jellyfishes and cephalopods; body undulations are used by 'worms' and chordates.
Swimming by body undulations	Swimming occurs by bending of the body by muscles acting against the hydrostatic skeleton turgor or antitelescopic notochord or vertebral column. Anguilliform swimming depends on reactive forces created to waves passing along the whole length of the body. Waves can be retrograde or direct. Carangiform swimming depends on thrust generated by the tail. Streamlining facilitates hydrodynamic efficiency. The median and lateral fins give stability. The heterocercal tail in sharks provides lift.
Related topics	Nonmuscular movement (C28) Buoyancy (C33) Locomotion: flight (C32)

Locomotory principles

All locomotion relies on an animal pushing against a fluid medium (water or air) or against the substratum (the surface of the ground or through soil, mud, sand, etc.). This generates a reaction, an equal and opposite thrust against the animal which pushes it forward (Newton's third law of motion!). Beating of flagella and cilia have already been considered (see Topic C28).

Changes in body shape

Methods used by animals which involve swimming by changes in body shape include:

- jellyfishes producing water jets by contracting muscle sheets in the bell;
- cephalopods (e.g. squids) producing water jets by expelling water from the mantle cavity through a funnel;
- body undulations in various invertebrate 'worms' (e.g. the nematode vinegar eel-worm *Anguillula* sp.), fishes (e.g. eel, mackerel) and marine mammals (e.g. whales).

Swimming by body undulations

The bending of the body is effected by waves of muscle contraction that alternate on the two sides of the body. The muscles act against the turgor of the coelom (e.g. in annelids), the turgor of body fluids maintained by a constraining cuticle (nematode worms) or the antitelescopic properties of the notochord (amphioxus) or vertebral column (fishes).

Anguilliform (eel-like) swimming

Here, the whole length (or almost the whole length) of the body undulates: the body length is equal to the wavelength or a whole-number multiple of the wavelength. In the eel, the waves are **retrograde** (backward moving). If components A and B (*Fig. 1*) are considered and reactive forces R and R′ to the movements of those components through the water are taken, the reaction can be resolved into propulsive thrust (P and P′) and lateral drag (L and L′). The lateral drag components cancel because they are in opposite directions, giving net propulsive thrust. (In the skate, anguilliform undulations of the large pectoral fin in a vertical plane propel this fish forwards.)

In the ragworm, the parapodial projections beat backwards as a **forward (direct) wave** moves from the tail to the head; the backward movements of the parapodia generate a net propulsive reaction.

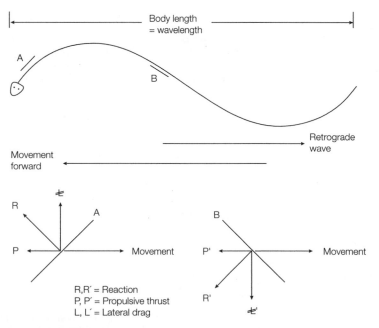

Fig. 1.   Anguilliform locomotion (as in an eel). For explanation of forces, see text.

Carangiform swimming

Here (as in a mackerel) only the tail provides thrust. As the tail sweeps from right to left (*Fig. 2*) there is a reaction R to its movement; momentum is lost by the time it reaches the center line and moves to the left of the fish so the reaction is very small. R can be resolved into propulsive thrust and lateral drag; the process is repeated in reverse as the tail then sweeps from left to right, generating a reaction R′. Resolving R and R′, the propulsive thrusts P and P′ summate (the backward thrust generated as the tail passes the mid-line is now very small because momentum is lost) and the lateral drag components, L and L′, cancel.

In dolphins, the tail beats in a horizontal plane (rather from side-to-side) to effect carangiform swimming. The head is large and is stable; streamlining of the body (especially in fast swimmers) provides hydrodynamic efficiency by ensuring a smooth, laminar flow (see Topic C32) of water over the body. Median

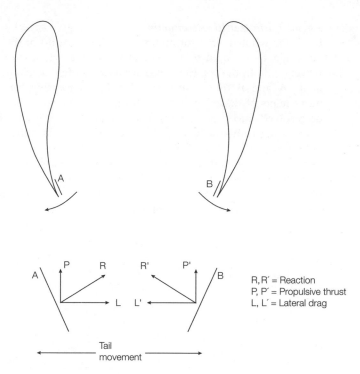

R, R' = Reaction
P, P' = Propulsive thrust
L, L' = Lateral drag

Fig. 2.   Carangiform locomotion (as in a mackerel). For explanation of forces, see text.

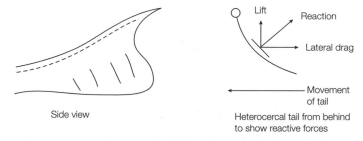

Side view

Heterocercal tail from behind
to show reactive forces

Fig. 3.   The shape of a heterocercal tail.

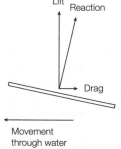

Fig. 4.   Forces acting
on a paired fin hydrofoil.

(dorsal and anal) and lateral (paired) fins assist stability and prevent pitching, yawing and rolling (pivoting around the x, y and z axes respectively).

In sharks, a **heterocercal tail** is thought to provide lift. The lower lobe of the tail hangs down like a curtain from the upper lobe into which the vertebral column is extended (Fig. 3). As the tail sweeps to the side, the lower lobe trails behind; reaction to the lobe can be resolved into lift and lateral drag. The use of the pectoral and pelvic fins as hydrofoils to provide lift at the anterior end (Fig. 4) counteracts the tendency of the shark to nose-dive as the result of the lift generated by the heterocercal tail.

C31 LOCOMOTION: TERRESTRIAL LOCOMOTION

Key Notes

Burrowing in annelids	Longitudinal and circular muscles in the body wall contract against the constant volume of coelomic fluids in the segments, changing the shape of the segments. Backward-moving waves of elongation and contraction, assisted by the grip of chaetae (setae) on the substratum, help to push the body forwards.
Hirudiform looping in leeches	The attachment of a leech to the substratum occurs by a sucker at one end of the body, followed by the extension of the body and the attachment of a sucker at the opposite end. The first sucker is released and the body contracts. This leads to a looping locomotion.
Walking and running with limbs	Walking is stable with the center of gravity above a triangle described by the three limbs on the ground. Running is not stable because the center of gravity is not above the shape described by the feet on the ground.
Tetrapod vertebrate legs	Legs have evolved from the fleshy fins of crossopterygian fishes. Diagonal progression using limbs as holdfasts is seen in belly-crawlers. An 'improved gait' swings the limbs under the body for more leverage and an increased velocity ratio.
Limbs as levers	Long limbs give a high velocity ratio [but less output force (leverage)] and so favor running; short limbs have more output force (greater leverage) but a lower velocity ratio, so are suitable for activities such as digging. There is a tendency for limbs to elongate from plantigrade through digitigrade to unguligrade gait to favor running. A graviportal system spreads weight in heavy animals.
Related topics	Phylum Annelida (A8) Skeletons (B3) Phylum Chordata (A14)

Burrowing in annelids

The locomotion of metamerically segmented annelid worms (e.g. earthworms) relies on the contraction of body wall muscles acting against the turgor of the coelom containing a constant volume of coelomic fluid. Circular muscle contraction elongates segments, and stretches longitudinal muscles. Contraction of longitudinal muscles conversely shortens and thickens the segments. Segmentation allows these two processes to occur simultaneously in different portional regions of the worm. Alternate waves of elongation and contraction move posteriorly along the length of the worm; the front of the body is elongated and pushed forward. The front then shortens and widens, chaetae (or setae, bristles) in the body wall anchor the front of the worm to the substratum, and the hind parts of the worm are pulled forward.

Hirudiform looping in leeches

Leeches possess anterior and posterior suckers; the leech can attach to the substratum by one sucker and then elongate the body by contracting circular muscles in the body segments. The other sucker is then placed on the substratum, the first sucker is released and the body shortens by contracting the longitudinal segmental muscles. The first sucker is then put down again and the process repeated, resulting in a looping, **hirudiform** (leech-like) loco-motion.

Walking and running with limbs

Arthropods and tetrapod vertebrates use limbs. During walking, the effective stroke of the limb on one side of the body alternates with that on the other side of the body. At most points during a stride, three limbs are on the ground and the center of gravity of the animal lies above a point within a triangle described by the three limbs: thus walking is a stable form of progression. In human standing, the center of gravity lies within a rectangle described by the toes and heels of the two feet on the ground; in walking, the body leans to the left as the right foot is lifted off the ground, so that the center of gravity lies above the left foot, and vice versa when the left foot is lifted off the ground.

When running, the gait changes and the center of gravity is not above a shape described by the feet on the ground. In human running, the center of gravity lies in front of and above the feet, the body is unstable and the runner is always trying to 'catch him/herself up'.

Tetrapod vertebrate legs

Tetrapod legs are derived from the fleshy fins of crossopterygian fishes (e.g. *Osteolepis* sp.): the **biserial archepterygian** pattern of bones in such fishes resembles the pattern of bones found in primitive labyrinthodont amphibians such as *Eryops* sp.

The earliest tetrapods were probably **belly-crawlers** in Devonian swamps. Their short, weak limbs were used as holdfasts, trunk movements being more important. Locomotion of this sort is described as **diagonal progression** (*Fig. 1*) and is used by newts and alligators today.

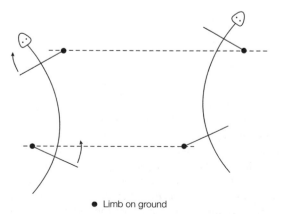

● Limb on ground

Fig. 1. Diagonal progression in a tetrapod.

To give a greater velocity ratio and leverage, there has been a tendency to swing the limbs under the body (as seen in 'dinosaur' groups, birds and most mammals, e.g. horses): this is known as **improved gait**; the distal ends of the forelimbs rotate below the elbow in a process called **pronation** so that the extremities face forwards.

Limbs as levers   Limbs can be seen as levers. The moment (force × distance) on one side of the fulcrum (pivot) is equal to the moment on the other side. In a first-order lever, for example a see-saw (*Fig. 2*), if F_i is the force put in and L_i is the length of the arm of the lever on the input side of the fulcrum, and F_o is the output force and L_o is the length of the arm of the lever on the output side, then:

$$F_i \times L_i = F_o \times L_o$$

In other words, if a small force is applied to a long arm of a lever, a greater force will be delivered if the output arm is short (think of using a crow-bar to lever up a heavy paving-stone). Thus F_o is large if L_i/L_o is large.

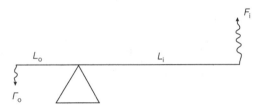

Fig. 2.   A first-order lever. L_i is the length of the input arm, L_o is the length of the output arm. F_i is the input force, F_o is the output force.

In **digging limbs**, as in a mole's forelimbs, L_i is large compared with L_o (*Fig. 3a*), giving a strong, powerful, levering leg. In a horse's leg, L_i is small compared with L_o (*Fig. 3b*): such long legs give high velocity ratios (the ratio of the velocity of the output arm of the lever to the velocity of the input arm) and favor **running**.

There has been a tendency for limbs to elongate to give a greater velocity ratio (*Fig. 4*):

- **plantigrade** gait has the whole foot on the ground (bear, human walking);
- **digitigrade** gait places the ball of the foot on the ground (cat, human running);
- **unguligrade** gait rests the weight on the tip of the digit, protected by a hoof or hooves (deer, horse, compare a woman ballerina on her 'points' (tip-toe) with blocks in her shoes).

Some unguligrade mammals (e.g. rhinocerous, elephant) are very heavy: the bones of the feet are broadened to allow less weight per unit cross-sectional area of bone and so spread the load. Such an arrangement is known as a **graviportal** system.

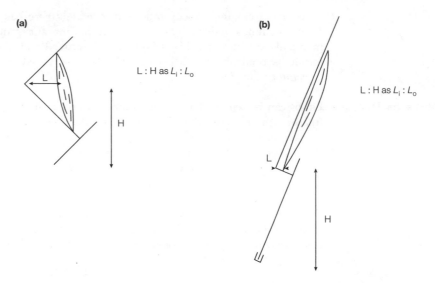

Fig. 3.   (a) A digging limb (e.g. mole forelimb); (b) a running limb (e.g. deer leg). L_1 and L_0 are the lengths of the input and output arms of the lever respectively. (a) L_1 is large; therefore, the ratio is low and the output force is high. (b) L_1 is small; therefore, the ratio is high, with speed favored. The output force is low.

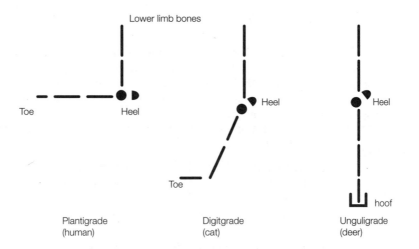

Fig. 4.   Comparison of limb length with gait. Plantigrade: the heel rests on the ground (human). Digitigrade: the ball of the foot rests on the ground (cat). Unguligrade: the tips of the toes rest on the ground, protected by hooves (deer).

C32 LOCOMOTION: FLIGHT

Key Notes

Aerial movement	Movement through the air can include parachuting, gliding and powered flight
Aerodynamic principles	An aerofoil, moving through a laminar flow of air, generates a reaction which can be resolved into lift and drag. The lift can be increased by increasing the angle of attack. An excessive angle of attack (destroying the laminar flow of air and creating turbulence) leads to stalling with loss of lift. Stalling is minimized by the use of a slot in the leading edge of the wing. Propulsion is effected by a downstroke of the wing, giving thrust.
Bird flight	Primary feathers provide a propulsive thrust, secondary feathers provide lift. Flight is aided by a light weight, an aerodynamic shape, homoiothermy, efficient gas exchange in a unidirectional lung, large, well-vascularized muscles and uric acid excretion. Lift is related to the area of wing surface, the angle of attack and the air-speed. Gliders use wings to provide lift, and gain forward motion through gravity. Static soarers have low wing loading ratios and low aspect ratios; dynamic soarers have high wing loading ratios and high aspect ratios. Flapping, powered flight uses thrust generated by a powerful wing downstroke; hoverers alternate downstrokes, one providing forward thrust (and lift), the next a counteracting backward thrust.
Insect flight	Insect wings are outgrowths of the thorax cuticle. Wing movements are controlled by indirect sternotergal and longitudinal muscles and by direct wing muscles.
Related topics	Phylum Arthropoda (A11) Locomotion: swimming (C30) Phylum Chordata (A14)

Aerial movement   Movement through the air can include:

- parachuting (e.g. 'flying' frogs, sugar-possums, flying squirrels);
- gliding (flying fishes, flying lizards, insects, birds);
- powered flight [insects, pterosaurs (extinct 'pterodactyls'), birds, bats].

In **parachuting animals**, an expanse of body surface (e.g. the webbing between the digits in flying frogs) provides resistance to air as the animal drops. In **gliders**, a plane surface (e.g. the expanded pectoral fin in flying fishes, a membrane of skin on extended ribs in the flying lizard *Draco* sp., the forelimb extended to form a wing in pterosaurs, birds and bats, folded integument in insect wings) acts as an aerofoil and provides lift (see below). In **powered flight**, the wing can also be used to provide a propulsive thrust force in reaction to its movement against the fluid air.

Aerodynamic   The aerodynamic principles of an aerofoil moving through air are similar to
principles   the hydrodynamic principles of a hydrofoil moving through water (see Topic
C30).

When a plane surface, an aerofoil, moves through air parallel to the air stream
it parts the layers of gas molecules making up the air: air pressure is equal
above and below the aerofoil. The flow of air is said to be **laminar**, the layers
of molecules, relative to each other, not being perturbed. If the aerofoil is
cambered (as in bird wings), movement of the aerofoil deflects air molecules
upwards above the aerofoil and pushes them against the lower surface of the
aerofoil. Thus the pressure is less on the upper, convex surface than on the
lower, concave surface and the air will have to move faster to cover the greater
distance across the upper surface. The reaction to the movement of the aerofoil is the aerodynamic force R (*Fig. 1*) which can be resolved into lift L and
drag D.

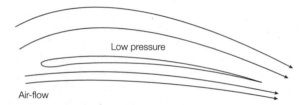

Fig. 1.   A cambered aerofoil passing through laminar air flow.

Lift L can be increased by tilting the aerofoil at an angle to the airstream.
This is called the angle of attack, θ (*Fig. 2*). However, as θ and L increase, so
D increases. At a certain point, the **laminar flow** of air over the wings is
destroyed and **turbulence** occurs: this is known as **stalling,** and lift is destroyed.
Stalling can be minimized by the provision of a slot in the leading edge of the
aerofoil (provided by the first digit, the **alula** or bastard wing in a bird's wing).
A wing with a sigmoid section can also reduce stalling (*Fig. 3*).

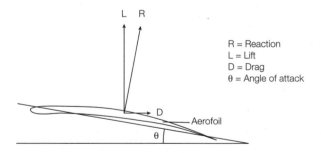

Fig. 2.   The angle of attack (θ).

Propulsion is provided by the flapping of the wing in bats and birds; the
downstroke or powerstroke produces most thrust whereas the upstroke encounters less air resistance through the position in which the wing is held.

Bird flight   Bird wings have **primary feathers** on the 'hand': these provide **propulsion**.
Secondary feathers on the forearm generate **lift** (*Fig. 4*). **Tertiary and contour**

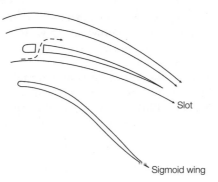

Fig. 3.   *The inclusion of a slot or a sigmoid wing section reduces stalling.*

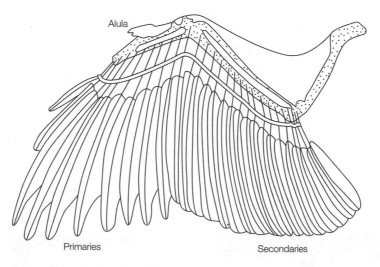

Fig. 4.   *The feathers of a bird's wing.*

feathers on the leading edge of the wing give a smooth, **aerodynamic shape**. The 'thumb' and its feathers is the **alula** forming a slot (see above and *Fig. 3*).

Flight in birds is aided by:

- light weight (air-sacs extend into **pneumatic bones**);
- overall aerodynamic shape;
- a homoiothermic metabolism with an efficient, unidirectional flow of air through the lungs (with a cross-current vasculature) facilitating efficient oxygen extraction from the air;
- large, well-vascularized muscles;
- detoxification of nitrogenous waste to uric acid, obviating the necessity to carry large volumes of water as a solvent.

Lift increases in proportion to the surface area of the wing (which can also be modified according to how much the wing is stretched or folded); lift also increases as the speed of flight increases (fast-flying birds such as swallows have relatively smaller wings than slower birds). For take-off and landing, the angle of attack can be increased to provide additional lift. In soaring birds, such as vultures, the spreading out of the feathers at the tips of the wing provides additional slots to reduce stalling (known as tip vortices). Tail feathers can also be fanned out to provide lift or to serve as a brake.

For **gliding**, the wings provide lift, and forward movement is provided by falling under the influence of gravity, although altitude can be gained through **static soaring** within a rising column or vortex of warm air (a thermal) or if air currents rise over hills. Birds which undertake static soaring (e.g. vultures) have wings with low aspect ratios (i.e. broad, short wings) (*Fig. 5a*):

$$\text{Aspect ratio} = \text{wing length} \div \text{wing width}$$

Such wings thus have a large surface area to provide adequate lift: the wing loading ratio is low:

$$\text{Wing loading ratio} = \text{weight of bird} \div \text{surface area of wing}$$

Tip vortices are reduced by slotting of the wing near the wing-tips.

In **dynamic soaring**, birds such as the wandering albatross use increases in air speed with increasing altitude over the ocean surface. Friction is highest just above the water, so the air speed is slowest. The bird commences by making a gravitational glide from a high altitude, downwind, gaining kinetic energy. Near the surface of the ocean, the bird turns upwind and uses its momentum to regain altitude, encountering progressively faster air speeds which give it additional lift until the bird regains its original altitude and the process is repeated. The albatross can remain in the air for many months, expending little energy. Unlike static soarers, dynamic soarers have wings with high wing loadings and high aspect ratios (*Fig. 5b*).

(a)   (b)

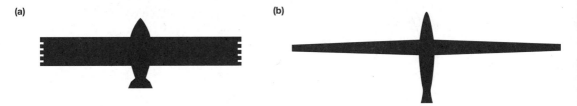

Fig. 5.   A stylized silhouette of (a) a static soaring bird and (b) a dynamic soaring bird.

Flapping (powered) flight uses energy. Wings move downwards and forwards during the downstroke (powered by the pectoralis muscles), inclined to the horizontal so that the leading edge is lower than the trailing edge. The reaction can be resolved into lift and drag, the lift having an upward and a forward (propulsive) component which reduces the drag component. On the upstroke (powered by the supracoracoideus muscle) the wing moves upwards and backwards in recovery, generating little reactive force (*Fig. 6*). The tail is used for support, balance and as a rudder.

Hovering flight: hoverers such as humming-birds hold the body vertical over or next to the flower from which they sip nectar. Lift is generated in a downstroke; after the recovery upstroke the wings rotate at the shoulder so that at the second downstroke the dorsal surface of the wing pushes downwards and backwards (i.e. opposite to the forwards push of the first downstroke).

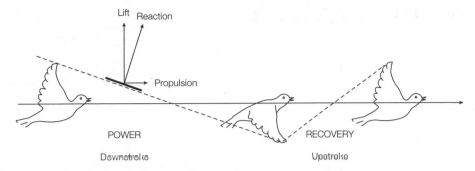

Fig. 6.   Wing strokes in powered flight. Redrawn from Elder and Truman (eds) (1980) Aspects of Animal Movement, *Society for Experimental Biology Seminar Series 5, with permission from Cambridge University Press.*

Insect flight

Insects are the only invertebrates which exhibit powered flight. Insect wings are outgrowths of the mesothorax and metathorax body wall and consist of cuticle. In beetles (Coleoptera) the mesothoracic wings are hardened elytra; in true flies (Diptera) the metathoracic wings form small balancers or halteres. Wings may be secondarily lost, as in fleas (Siphonaptera).

Wing movements are controlled by thoracic flight muscles: direct flight muscles are attached to the wings themselves, whereas indirect flight muscles cause wing movements by altering the thorax shape. The wing is hinged at the tergum (upper casing) of the thorax and on a pleural process (*Fig. 7*). Indirect sternotergal muscles pull down the tergum and effect an upstroke of the wing; the downstroke in small insects (e.g. houseflies) occurs when the sternotergal muscles relax and longitudinal muscles in the thorax arch the tergum again. In some insects (e.g. grasshoppers), this process using indirect muscles is assisted by muscles that attach directly to the base of the wing.

Forward thrust is effected through interplay of contractions of direct and indirect muscles: the indirect muscles contract in rhythm alternately, the direct muscles altering the angle of the wings so that the wings act as aerofoils providing lift during both downstroke and upstroke, twisting the leading edge of the wing down in downstroke and up in upstroke. This gives a figure-of-eight movement.

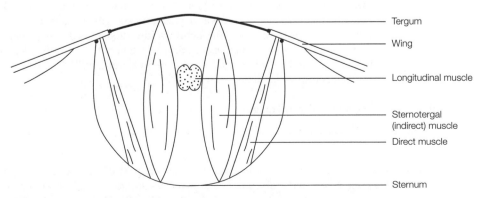

Fig. 7.   Cross-section of an insect thorax.

C33 BUOYANCY

Key Notes

General principles	Animals with neutral buoyancy have the same density as the water in which they live, this saves energy. The density can be lowered by replacing heavy substances or by replacing heavy ions by lighter ions, by making the animal hypotonic to the medium, by use of light fats and oils, and by the use of gas floats.
Reduction of heavy substances	The removal of calcium salts leads to weight loss (e.g. loss of shells).
Ion replacement	Cranchid squids retain light ammonium ions and exclude heavy sulfate ions; jellyfishes also exclude sulfate ions.
Hypotonicity	Teleosts are hypotonic to sea water although this may not aid buoyancy significantly.
Fats and oils	Swimming elasmobranchs (e.g. sharks) have large livers with squalene, a low-density oil. Sperm whales use spermaceti oil in head cavities for buoyancy.
Gas floats	Soft-walled gas floats (e.g. teleost fish) contain gas at the same pressure as the hydrostatic pressure outside; a means of compensation is required as the animal moves up or down. Rigid-walled gas floats (e.g. cuttlefish) allow freer vertical movement – these are stronger but heavier.
Invertebrate gas floats	In *Physalia* sp. (Portuguese man-o'-war), a balloon-like float gives buoyancy and acts as a surface sail. Cuttlefishes have a calcium carbonate cuttlebone which is full of chambers filled with gas and/or fluid. This fluid can be withdrawn or secreted to adjust buoyancy.
Fish swim-bladders	The swim-bladder is homologous to a lung. Physostome fishes have bladders with a connection to the esophagus; they gulp air on the surface to fill them. Physoclist fishes have closed bladders. Gas is secreted into the swim-bladder by a gas gland as lactic acid drives oxygen off hemoglobin against considerable gas pressures in the swim-bladder (Root effect). Gas is conserved in the countercurrent exchanger/multiplier system of the rete mirabile, which is characterized by a large capillary surface area, thin capillary walls and long capillaries (especially in deep-sea fishes).
Related topic	Locomotion: swimming (C30)

General principles If a swimming animal is heavier than water, part of its energy is expended in keeping afloat and only part for foward locomotion. If an animal has **neutral buoyancy** (i.e. the density, ρ, of the animal = the density, ρ, of water) this is energy efficient (positive buoyancy favors floating, whereas negative buoyancy favors sinking). This consideration is important for larger animals. Smaller animals have larger surface areas relative to volume, and surface extensions can reduce their sinking rates.

Lowering ρ is the easiest way to aid buoyancy and is the only method which avoids active swimming. There are five possible ways:

(1) reduction of amounts of heavy substances, e.g. calcium carbonate, calcium phosphate;
(2) replacement of heavy ions (e.g. Mg^{2+}, SO_4^{2-}) by lighter ions such as Na^+, Cl^- or even NH_4^+ or H^+;
(3) making the animal hypotonic to the water;
(4) increasing amounts of light fats or oils;
(5) using a gas float.

Removal of calcium salts can contribute greatly to weight loss.

Reduction of heavy substances

$$\rho \text{ aragonite (a form of calcium carbonate)} = 2.9 \text{ g cm}^{-3}$$

$$\rho \text{ chitin} = 1.2 \text{ g cm}^{-3}$$

Nudibranch gastropods lack shells and swim; bottom-dwelling opisthobranch gastropods possess shells. Swimming jellyfishes lack heavy skeletons compared with sessile corals. Loss of a skeleton makes the animal more vulnerable to predation and removes a means of support. Many swimming animals (e.g. blue crabs) retain their exoskeletons, even at the cost of more energy expenditure when swimming. The squid has replaced the heavier calcium carbonate shell with a chitinous 'pen' while the octopus has lost its shell totally.

Ion replacement

Cranchid squids have a coelom filled with a fluid with $\rho = 1.011$ g cm^{-3}: the coelom comprises about 67% of the body volume and if it is punctured the squid sinks. The coelomic fluid is isotonic to sea water with an ammonium ion concentration of 480 milliequivalents per liter as compared with a sodium ion concentration of 80 milliequivalents per liter. The ammonium ions are trapped from protein metabolism. The anion is chloride, the heavier sulfate ions being excluded. The pH is 5.2, the result of a sodium pump leaving an excess of anions and consequent high proton concentration. (The jellyfish *Aurelia* sp. also excludes heavy sulfate ions.)

Hypotonicity

Removal of ions without replacing them leads to osmotic problems: teleost fishes are hypotonic but any weight saving is counteracted by muscles and bone. A few deep-sea teleosts have reduced skeletons and this may assist buoyancy.

Fats and oils

Elasmobranch fishes have large, oily livers making up nearly 20% of the body weight (the equivalent figure for teleosts is 2%). The shark *Etmopterus spinax* has a liver making up 17% of the body weight: 75% of the liver is oil. Half of this oil is squalene, $C_{30}H_{50}$, $\rho = 0.86$ g cm^{-3}; most oils have densities between 0.90 and 0.92 g cm^{-3}. Bottom-living skates and rays have smaller livers with less fat. Some abyssal teleosts have swim-bladders full of cholesterol ($\rho = 1.067$ g cm^{-3}); its role in buoyancy is dubious.

Sperm whales have a spermaceti organ in the head. When the whale dives the oil in the organ is cooled and becomes denser, facilitating neutral buoyancy and hovering at one depth while it feeds on squid. On surfacing, warm blood is pumped into the organ: the oil warms and becomes less dense, helping the whale to surface.

Gas floats

Gas has a low density so that even a small volume can assist buoyancy. The presence of rigid *or* flexible walls on a gas float is important.

Soft-walled gas floats
As the float descends, the pressure on the outside increases and is matched by that inside. This pressure increases by 100 kPa (1 atmosphere) for every 10 m depth. Gas must be secreted against this pressure, so the amount of gas secreted to fill a given volume = n × the amount of gas secreted on the surface, when the hydrostatic pressure outside is 100 n kPa [i.e. if the outside pressure is 500 kPa (at 40 m depth) 5 × as much gas must be secreted]; there will also be a tendency for gas to diffuse out at this pressure. The gas in the float must be obtained from sea water where the oxygen tension is always 20 kPa, the nitrogen tension 80 kPa. As the animal moves up and down, the gas volume changes, together with its density: thus the animal will have neutral buoyancy at only one depth. If the animal descends, the weight (not the mass) of the animal rises; if the animal ascends, the gas expands, risking rupture of the float.

Rigid-walled gas floats
A rigid wall allow more vertical movement, although the walls must be stronger (and thus heavier).

Invertebrate gas floats

In the siphonophore cnidarian *Physalia* sp., the Portuguese man-o'-war, there is a balloon-like float containing 85% nitrogen and 15% carbon monoxide; the float, about 10 cm long, acts as a sail whereby the *Physalia* colony is blown around on the ocean surface. The gas is held at approximately 100 kPa pressure: L-serine acts as the substrate in the synthesis of carbon monoxide. The float has a valve to allow release of excess gas. Rigid floats in invertebrates permit ascent and descent in the water (N.B. *Physalia* remains on the surface.)

In the cephalopod cuttlefish *Sepia* sp., 9.3% of the body volume is made up of the cuttlebone (ρ = 0.57–0.64 g cm^{-3}). The bone is laminar, comprising thin layers of chitin-reinforced calcium carbonate. The lamellae are held 0.7 mm apart by pillars of the same material. The spaces between the lamellae are filled with (mainly) nitrogen at about 80 kPa. In the older, more posterior chambers of the bone there is liquid containing NaCl hypotonic to the sea water and body fluids: its weight helps to keep the tail horizontal when swimming. The higher osmotic pressure outside the cuttlebone tends to withdraw water from the bone while hydrostatic pressure tends to push water back in, establishing a balance. In sea water, the cuttlebone gives a net lift of 4% of the cuttlefish's weight and so approximately balances the excess weight of the rest of the cephalopod. Despite its bulk, the cuttlebone has the advantage as a buoyancy device of being more or less independent of depth. Density *can* be reduced by pumping (by active transport) ions out of the cuttlebone, with water following by osmosis (during the night, allowing ascent), or the process may be reversed, making it more dense, during the daytime.

In *Nautilus* sp., new float chambers are added to the external shell as the cephalopod grows; initially the new chambers are filled with NaCl solution. Sodium ions are actively transported out when the shell is strong enough (the pump is located in a strand of tissue, the siphuncle, passing through the chambers), water following passively by osmosis. Gas then diffuses in: nitrogen at 80 kPa and oxygen at 10 kPa (total pressure about 90 kPa). Freedom of vertical movement is only constrained by the ability of the shell to withstand pressures.

Fish swim-bladders

Many teleosts have a swim-bladder, its presence or absence reflecting life-habit, not taxonomic position. Some bottom-dwellers lack swim-bladders, having no need for buoyancy; a few powerful, fast-swimming fishes lack them too (e.g. the mackerel) where the swim-bladder may be disadvantageous, placing constraints on rapid vertical movements. (An analogy between an airship with positive buoyancy and low-power engines and an airplane with negative buoyancy but powerful engines to give enough thrust to provide sufficient lift is appropriate.)

Typically, the swim-bladder is an oval sac, lying below the vertebral column in the abdomen. It is the homolog of the lung. In marine teleosts it comprises 5% of the body volume, in freshwater species 7%. $\rho = 1.07$ g cm^{-3} for a bladderless teleost, $\rho = 1.02$ g cm^{-3} for a fish with a swim-bladder.

The teleost is only at neutral buoyancy at one depth: descent results in a decrease in buoyancy with the necessity to work harder to compensate. Ascent endangers the fish through the swim-bladder bursting.

In **physostome** fishes, there is a connection between the swim-bladder and the esophagus, and the bladder can be filled by gulping air. In deeper-sea fishes, such gulping would be impractical and the connection has been lost. This is known as a **physoclist** condition. In physoclist flishes the gases must come from the blood and must be secreted at the pressure at which the fishes live. Bladdered fishes have been found at 4000 m depth, that is with gas at a pressure of 40 000 kPa (400 atmospheres).

The bladder is filled with a gas mixture rich in oxygen (in the whitefish, for reasons not fully known, the gas is mainly nitrogen).

Gas retention

A gas gland (see below) is located in the swim-bladder wall. If this was supplied with arterial blood directly from the gills the venous blood leaving the gland would tend to dissolve the swim-bladder gases and carry them away.

The gas gland receives blood which has first passed through the **rete mirabile** ('wonderful net'), Before the artery reaches the gas gland it splits into large numbers of straight, parallel capillaries which reunite before reaching the gas gland. The vein draining the gas gland does the same, parallel venous capillaries running among and alongside the arterial capillaries. This arrangement (*Fig. 1*) permits a **countercurrent exchange** system for dissolved gases. If the bladder gas is at 1000 kPa, venous blood draining the gland contains gas also at this pressure; in the rete mirabile the gas is lost to the blood contained in the inflowing, arterial capillaries. On leaving the rete the venous blood has no more gas than does the arterial blood flowing into the rete, and gas is thus conserved.

Characteristics of the rete mirabile are:

- large surface area: many capillaries (area about 10^4 mm^2 in an eel);
- short diffusion distance (1 μm thick capillary walls);
- long capillaries, several millimeters long: the length is proportional to the pressure in the swim-bladder. The deep-sea *Bassozetus* sp. has 25 mm long capillaries (compare muscle capillaries of about 0.5 mm length).

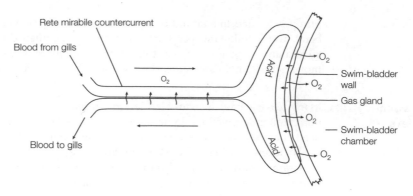

Fig. 1.   A teleost swim-bladder and rete mirabile.

Gas secretion

Assume a fish needs to secrete oxygen against a pressure in the swim-bladder of 1000 kPa (10 atmospheres). The oxygen pressure in arterial blood is 20 kPa. If oxygen is deposited in the swim-bladder the venous blood will contain less oxygen when it leaves the rete mirabile, the difference being the oxygen secreted into the swim-bladder. The gas gland produces lactic acid which enters the blood, reducing the oxygen affinity and capacity of the hemoglobin: this is known as the **Root effect** (*Fig. 2*) (the Root effect is a type of Bohr effect; see Topic C4), and oxygen is released into the swim-bladder. The oxygen pressure (tension) in the venous blood leaving the gas gland is increased as the gas dissolves. In the rete, the oxygen diffuses into the inflowing capillaries and the gas is conserved in the system (*Fig. 1*). The lactic acid also remains in the rete system because of the countercurrent, further enhancing the process. Oxygen pressures build up so that there is a **countercurrent multiplier** effect (compare the loop of Henle in the mammalian kidney; see Topic C17).

Lactic acid not only exerts a Root effect, it also 'salts out' blood gases, including nitrogen, explaining its presence in the swim-bladder of many species (e.g. whitefish).

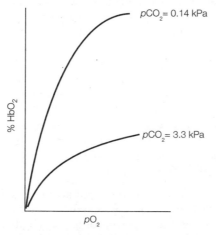

Fig. 2.   The Root effect: gas secretion. Oxygen–hemoglobin curves at different pCO$_2$ values. (lactate has a similar effect).

D1 REPRODUCTION, GAMETOGENESIS AND FERTILIZATION

Key Notes

Asexual reproduction	Fission, fragmentation and budding facilitate asexual reproduction resulting in genetically identical new animals or colonies. This allows rapid replication (or regeneration) but not genetic variability.
Sexual reproduction	This results in shuffling within the gene pool and genetic variation but is energetically expensive. Haploid gametes (sperm and eggs) are produced by meiosis: fusion at fertilization results in a zygote. Evolutionary origins may be linked to the DNA repair facility. Protoctistan conjugation may not be linked to reproduction. Most animals are dioecious (separate sexes) but hermaphroditism is found in many species, especially sessile forms.
Parthenogenesis	Unfertilized eggs can be stimulated to develop. This is naturally found in some species where males are unknown (e.g. some lizards) or where several parthenogenetic generations of females are followed by a generation with sexually reproducing males and females (e.g. aphids).
Gametogenesis in mammals	*Sperm development*: sperm is produced in the testis by spermatogenesis, frequently seasonally. Spermatogonia and Sertoli cells are found in the seminiferous tubules of the testis. At puberty, these develop under the influence of gonadotropins. The mitotic division of spermatogonia to form primary spermatocytes and subsequent meiosis gives secondary spermatocytes and then spermatids which mature without further division into sperm. Final maturation and storage occur in the epididymis. Semen is formed as secretions are added to sperm suspension as it passes along the male gonoduct. *Egg development*: Eggs develop by oogenesis. Germ cells (oogonia) expand in numbers mitotically and become primary oocytes whose development is arrested in prophase of the first meiotic division. After puberty (in women), meiosis is completed to form a secondary oocyte and then an ootid which matures into an egg (ovum).
Fertilization	External fertilization occurs when eggs and sperm are shed into water. Internal fertilization occurs when the egg is retained in the female until after fertilization. External fertilization is aided by processes to bring the gametes together; internal fertilization is aided by the release of sperm near the eggs and by the capacity of the female (in some species) to store sperm. Internal fertilization involves copulation and possible use of a sperm package (spermatophore). Fertilization entails recognition events such as species-specific chemical attractants, the acrosome reaction (whereby the sperm can penetrate the egg membranes), and sperm-egg binding. Polyspermy is

prevented by membrane polarity reversal, the cortical reaction and formation of a fertilization envelope in some species.

Reproductive synchrony

This is facilitated by environmental cues, courtship behavior or, in sessile forms, the presence of gametes of opposite sex.

Egg laying

Unprotected eggs have a poor surival rate, so large numbers of eggs are produced. Enclosure of eggs in an envelope or shell enhances survival, which is essential on land. Fertilization occurs before the envelope or shell is formed, or through the micropyle of an insect egg.

Reproductive tracts (gonoducts)

Simple gonoducts are found in animals relying on external fertilization. Gonoducts are modified for copulation, gamete storage and egg envelope production in forms using internal fertilization.
Male gonoduct: in male fishes and amphibians, sperm leave via the opisthonephric kidney and its duct; in amniotes the metanephric kidney has its own duct. The opisthonephros becomes the epididymis and the opisthonephric duct becomes the vas deferens.
Female gonoduct: in female vertebrates, the oviduct transports eggs. Glands may be present to secrete egg mucilage, albumin or shell. Viviparity is associated with the development of a uterus from the hind part of the oviduct.

Related topics

Nitrogenous excretion (C17)   Puberty in humans (D5)
Hormones (C22)   Menstrual cycle in women (D6)
Development and birth (D2)

Asexual reproduction

In asexual reproduction the parent divides, fragments or buds to give rise to a varying number of genetically identical offspring. This 'natural cloning' is the result of mitotic cell division. The simplest form of asexual reproduction is seen in **fission** where the parent divides into two (binary fission) or more parts, each becoming an independent organism. Among the Protoctista, binary fission occurs in *Amoeba* spp.; in metazoan animals, platyhelminth flatworms exhibit reproduction by fission. Budding is seen in Cnidaria (e.g. *Hydra* spp.) and Urochordata (e.g. *Pyrosoma* sp.): buds on the parent's body differentiate and break free to form new organisms or remain attached to form components in a colony.

Lizards can regenerate new tails if the original is lost, starfishes can regenerate a whole new organism from an isolated arm. This regeneration is related to asexual reproduction.

Many parasitic worms reproduce asexually during their larval stages. It could be argued that the formation of monozygotic, identical twins in vertebrates, including humans, is a form of asexual reproduction.

Asexual reproduction results in the production of large numbers of progeny very speedily. However, it does not shuffle the gene pool (only mitosis is involved in cell replication with none of the opportunity for recombination of genetic material which is associated with meiosis) and so variation (on which natural selection can work) does not occur except by genetic mutations. Consequently, asexual reproduction is advantageous in a constant environment but disadvantageous if the environment is changing.

Sexual reproduction

In sexual reproduction, the gene pool *is* shuffled by the processes of meiosis and the exchange of genetic material between two animals. In some species of *Paramecium*, two individual ciliophorans may come together (**conjugation**) and exchange genetic material before separating: here no reproduction has occurred. This phenomenon may be associated with DNA repair (see below). In many other protoctistans and in animals, genetic recombination is linked with reproduction.

Sexual reproduction involves the formation, by **meiosis**, of **gametes**, that is the female **egg** (ovum) and the male **sperm** (spermatozoon). Each gamete has the haploid (half) number of chromosomes for the species. At fertilization, the egg and sperm fuse to form a **zygote** and the diploid (complete) number of chromosomes is restored.

Sexual reproduction is energetically very expensive because large numbers of gametes must be generated, of which only a few are used to form zygotes which grow to reproductive maturity. Even in birds and mammals where few eggs are produced, large numbers of sperm are generated and the female must expend considerable resources on the nurture of the developing embryos and infants. The advantage of sexual reproduction is the generation of variation upon which natural selection can act: such variation may be particularly advantageous in changing environments.

The evolution of sexual reproduction may be linked to another advantage associated with it: the opportunity for the repair, during the first prophase of meiosis, of damaged DNA. (Incorrect nucleotides on a DNA strand can be detected, deleted and replaced with a correct nucleotide, using the complementary DNA strand as a template.) The increase in genetic variability is a byproduct of this process, further amplified by linking it to reproduction.

In most animal species the male and female sexes (genders) are distinct (**dioecious**), although **hermaphroditism** is found in many species, especially those which are sessile. Hermaphroditism is advantageous in that gametes from any one individual stand a chance of meeting gametes of the opposite sex from any other, neighboring individual rather than from just a proportion of such individuals. Asynchrony in the production of eggs and sperm (e.g. through the testes developing before the ovaries, i.e. **protandry**, as in some oyster species) or other arrangements will be necessary to avoid self-fertilization, although some tapeworms do self-fertilize.

Parthenogenesis

Unfertilized eggs can be stimulated into development, a process called parthenogenesis ('virgin origin'). Artificial parthenogenesis can be induced in some species (e.g. the toad *Xenopus laevis*) by temperature, pH or mechanical shock to the egg: the resulting animal is often small and sterile. Natural parthenogenesis occurs in many species. In some rotifers and in some lizard species, males are not known. The eggs are either diploid, or the haploid number of chromosomes doubles as the egg starts its development. In some species of *Daphnia* (a crustacean water-'flea') and aphids (hemipteran insects), parthenogenetically produced females occur for several generations, followed by haploid males which mate with haploid females.

Gametogenesis in mammals

Sperm development

Sperm are produced in the testes by a process of **spermatogenesis**. In some species (e.g. humans), the testes remain the same size, regardless of season; in others the testes enlarge in, and spermatogenesis is confined to, a breeding season. **Germ cells** are recognizable in the embryo: they colonize the gonad. In male humans, genes on the Y chromosome determine the development of testes and subsequent male morphology. (In the absence of these genes, the undifferentiated gonad develops into an ovary and the embryo develops into a female.)

In human male embryos the germ cells multiply in the highly coiled **seminiferous tubules** of the testis by mitosis until 5–6 months of gestation; they then become quiescent until puberty. The walls of the seminiferous tubules comprise clusters of sperm-forming cells, **spermatogonia**, interspersed with supporting **Sertoli cells** (nurse cells). At puberty, gonadotropins (FSH and LH) from the anterior pituitary initiate spermatogenesis (and testosterone production by Leydig cells in the testis, leading to the development of secondary sexual characteristics).

Spermatogonia undergo further mitosis (throughout adult life) to yield more spermatogonia and also **primary spermatocytes** (*Fig. 1*) which enter meiosis. The

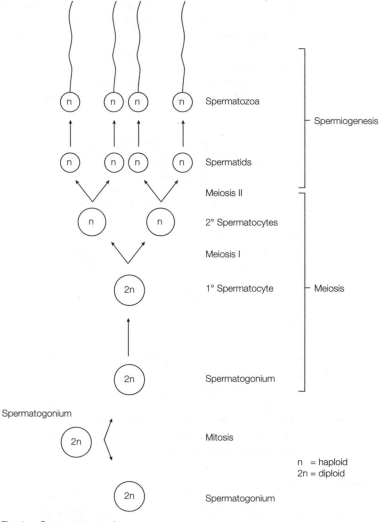

Fig..1. Spermatogenesis.

first (reduction) division of meiosis gives **secondary spermatocytes**, and the second meiotic division gives **spermatids** which further differentiate (without further division) into **sperm** (spermatozoa). The nucleus of the spermatid becomes condensed and most of the cytoplasm is phagocytosed by the Sertoli cells; secretory granules accumulate at the anterior end of the cells to form the acrosome (a modified lysosome). The two centrioles of the spermatid relocate behind the nucleus, the hindmost one forming the axial filament of the sperm tail which has the characteristic 9 + 2 structure of a flagellum (see Topic C28). A sheath may thicken and strengthen the proximal end of the tail. Mitochondria are found in the middle piece of the sperm, behind the head (*Fig. 2*). Sperm in arthropods (e.g. lobsters) may lack a tail and move by pseudopodia.

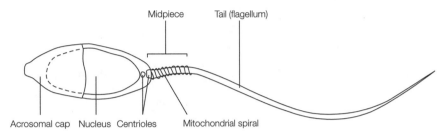

Fig. 2.   A typical mammalian spermatozoon.

The sperm are released into the seminiferous tubules and then to the epi-didymis where maturation is completed. In men the time lapse between the primary spermatocyte and sperm release is about 74 days, maturation in the epididymis taking about 12 days. Ejaculation is via the male gonoduct – the vas deferens and the urethra. The sperm suspension is mixed with nutritional and lubricatory secretions from various glands in the male reproductive tract (e.g. the prostate, Bowman's gland, Cowper's gland) to form semen. In a normal man, a 5 ml ejaculate of semen contains about 3.5×10^8 sperm. In most mammals, sper-matogenesis is enhanced by lowered temperatures, as will occur when the testes are contained within the scrotum, outside the abdominal cavity: in some species the testes are withdrawn into the abdomen outside of the breeding season.

Egg development
Eggs develop by **oogenesis**. Ovary size and activity may again be dependent on season. In many fishes, the ovaries are very large because of the large number of eggs produced; in birds and reptiles the production of yolky eggs also results in large ovaries. Germ cells (**oogonia**) undergoing mitotic expansion are found in the cortex of the fetal ovary. These cease mitosis and become **primary oocytes** between 2 and 9 months of gestation (in humans). By birth, all the germ cells are primary oocytes whose first meiotic division has been arrested at the end of prophase. Under the influence of gonadotropins following puberty (see Topics D5 and D7), typically one primary oocyte per month (in women) completes its first meiotic division to form a **secondary oocyte** (containing most of the cytoplasm) and a small first polar body. The second meiotic division to form an **ootid** which matures to form the egg (ovum) (and a second polar body) only occurs following fertilization (*Fig. 3*). Fusion of the egg and sperm nuclei produce the zygote. The polar bodies degenerate.

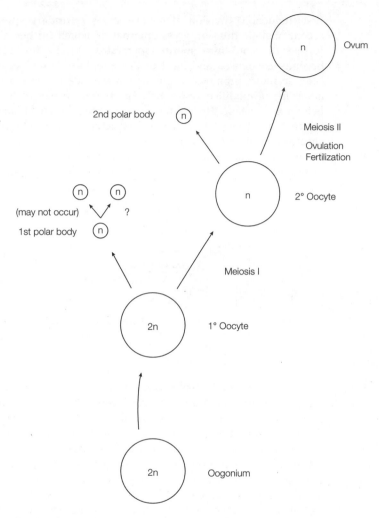

Fig. 3.   Oogenesis.

Fertilization

The gametes usually leave the body via rupture of the body wall in primitive species or, more usually, by **gonoducts**, the sperm duct or the oviduct. In **external fertilization** the eggs are shed into the water where they meet swimming sperm. In **internal fertilization** (associated with viviparity and/or terrestrial life) the egg is retained within the female's body until after fertilization (as in reptiles) or even until the embryo has developed into a more or less mature infant (as in placental mammals). The sperm are introduced into the female's body, often using an intromittent organ (e.g. the mammalian penis) – see below.

External fertilization is aided by:

● water currents carrying gametes (particularly important for sessile or sedentary animals);
● sperm carried in feeding and ventilatory currents in water;

- dense populations of individuals (e.g. mussels);
- swarming (e.g. palolo annelid worms);
- physical contact between individuals (e.g. frogs);
- production of vary large numbers of gametes (100×10^6 eggs in oysters).

Internal fertilzation is aided by:

- the release of sperm near to the eggs, permitting the production of fewer gametes and a saving of resources;
- the capacity for many females to store sperm after mating.

Internal fertilization involves **copulation** [with anatomical modifications such as a male intromittent organ (e.g. a penis) and a female receiving chamber (e.g. a vagina)] in order to transfer the sperm, usually suspended in fluid semen, although a **spermatophore** package is deposited in some species (e.g. the axolotl, a urodele amphibian). Some flatworms use a stylet on the penis to inject sperm beneath the partner's skin, and several snail species 'fire' a calcareous dart to stimulate the partner.

Gamete lifespan is usually short, and fertilization must occur in many species within about 40 hours. Sperm and eggs of the same species 'recognize' each other and a reaction is initiated. Recognition can be facilitated by the release of chemical attractants into the water from the egg, specific to the species.

The **acrosome reaction** (elucidated in sea-urchins) is the result of the recognition event:

- when sperm touches the jelly surrounding the egg, a polysaccharide in the jelly alters the permeability of the acrosomal membrane and Ca^{2+} ions from the surrounding fluid enter;
- acrosomal enzymes are released which digest a way through the jelly;
- actin molecules from the acrosome polymerize, forming an acrosomal process extending to the vitelline envelope overlying the egg plasma membrane: the acrosomal process has signal molecules which link to specific receptors on the egg membranes of compatible but not of incompatible eggs;
- the vitelline membrane breaks down;
- the egg extends a fertilization cone towards the sperm;
- sperm and egg plasma membranes unite: the sperm nucleus and proximal centriole are taken into the egg.

Polyspermy (entrance of more than one sperm nucleus) is prevented by:

- rapid entry of Na^+ ions, reversing the membrane polarity and preventing additional sperm binding, immediately after the entrance of the first sperm;
- a **cortical reaction** spreading from the entry site of the first sperm;
- the formation of a fertilization envelope (following release of Ca^{2+} ions and mucopolysaccharides and an osmotic inflow of water) between the plasma membrane and the vitelline envelope.

In the egg, the sperm nucleus swells to form a male pronucleus that migrates towards the female pronucleus of the egg. The pronuclei fuse in many species, or chromosomes join the mitotic spindle leading to the first division of the zygote, one centriole deriving from the sperm, the other from the egg. Calcium ion increase (which led to the cortical reaction above) activates the egg cytoplasm into metabolic activity in preparation for further growth and division.

Reproductive synchrony

For fertilization, it is important that gametes are produced and then released at the same time. In most animals these events are seasonal (although not in humans). Environmental cues such as temperature, day length or tidal cycles trigger nervous or hormonal initiation of reproduction-related events such as gamete production or release. In many sessile animals gamete release into the water by one sex will stimulate gamete release by the opposite sex (e.g. in sponges); courtship behavior (e.g. in some polychaete annelids) can also act as a trigger.

Egg laying

In many aquatic animals, eggs merely covered by a membrane are shed into the water: survival rates are very poor so large numbers of eggs are produced. Enclosure of the egg (or groups of eggs) in a horny, leathery or gelatinous envelope enhances survival: they may further be deposited on the substratum or attached to a rock or to vegetation. Fertilization normally must occur internally before the envelope is formed (in insects the egg-case has a micropyle pore through which the sperm can enter). The terrestrial reptiles, birds and monotreme mammals have waterproof leathery or calcareous egg shells.

Reproductive tracts (gonoducts)

Animals which use external fertilization (e.g. starfishes) have simple, tube-like gonoducts to allow transport of gametes to gonopores for release to the outside.

Animals which use internal fertilization may have male gonoduct vesicles for sperm storage prior to copulation, glands associated with the preparation of semen and a terminal portion which functions as a penis. The female gonoduct may have a chamber to receive the penis and possibly a sperm storage chamber. The female gonoduct (oviduct) may be extensible to allow the passage of large eggs, and may have glands associated with it to secrete albumin, egg envelopes or shells, or adhesive mucilage.

The vertebrate male gonoduct
In most fishes and amphibians the sperm is carried from the testis seminiferous tubules through **vasa efferentia** to the anterior of the opisthonephric kidney (see Topic C17): the sperm leave via the opisthonephric duct (which also serves, not simultaneously, to drain urine from the kidney).

In amniotes (reptiles, birds and mammals), a separate, nonsegmental metanephric kidney has evolved. This kidney is drained by its own ureter. The opisthonephric duct of fishes and amphibians is now utilized solely to transport sperm and has become the **vas deferens**: the former opisthonephros has become the **epididymis**, the vasa efferentia being reduced to a tiny **rete testis**. Sperm are stored in the epididymis and the vas deferens, where they mature. Final acquisition of fertilizing capacity (capacitation) is only achieved after a period in the female tract in mammals. Birds lack a penis; the male brings his cloaca against the female's cloaca during mating to facilitate internal fertilization; in mammals the penis become erect when aroused erotically, due to arterial dilation and consequent turgor of vascular cavities within the organ, maintained by restriction of venous return. Associated with the male gonoduct are seminal vesicles and glands which secrete components of seminal fluid.

The vertebrate female gonoduct (oviduct)
In fishes and most amphibians where fertilization is external, eggs are shed into the coelom from the ovary; the eggs enter the oviducts through **ostia** and pass down the oviducts where layers of jelly may be secreted around the eggs

(e.g. frogspawn). The eggs may be temporarily stored in **ovisacs**, and are then expelled via the cloaca. The paired oviducts (or **Müllerian ducts**) are found in all female vertebrates: vestigial oviducts may be present in males. Unlike the situation in males, the opisthonephric kidney and its duct is not recruited for gamete transport.

In reptiles and birds (which use internal fertilization), a large egg is deposited (except in a few viviparous reptiles such as some snakes). The oviductal glands are large and secrete the albumin and shell. In birds, one of the paired oviducts is usually lost.

Viviparity is seen in marsupial and placental mammals, together with a few fishes, amphibians and reptiles (never in birds). Membranes surrounding the embryo come into close contact with a region of the oviduct (the **uterus** or womb in mammals) to form a **placenta** through which exchange of nutrients, gases, hormones, certain antibodies, waste products, etc. takes place. In placental mammals, the ostium at the top of the oviduct assumes a funnel-like shape and partially surrounds the ovary. When an egg is released from the ovary it is usually wafted into the oviduct by ciliary currents and carried down the anterior part of the oviduct, the convoluted **Fallopian tube**, aided by muscular contractions. The lower regions of the oviduct form the thick-walled, muscular uterus: in some species of mammals (e.g. humans) the oviducts fuse here. The neck of the uterus, the **cervix**, has a sphincter-like morphology, and is separated from the **vagina** formed from the most posterior part of the oviduct and the ventral parts of the cloaca: the vagina forms a receptacle to receive the penis during copulation.

D2 DEVELOPMENT AND BIRTH

Key Notes

Egg types	Egg types can be characterized as isolecithal, telolecithal or centrolecithal according to the amount and localization of the yolk.
Cleavage patterns	The zygote undergoes cleavage to give a morula of blastomeres which develops into a hollow blastula surrounding a blastocoel cavity. In holoblastic cleavage the whole blastula cleaves; in meroblastic cleavage the process is incomplete. Radial cleavage is characteristic of deuterostome phyla; spiral cleavage is characteristic of protostome phyla.
Mosaic and regulatory control	*Determinate cleavage*: this is found in animals demonstrating spiral cleavage (e.g. molluscs) and a few showing radial cleavage (e.g. sea-squirts): the prospective fates of blastomeres are rigidly determined, giving mosaic development. *Indeterminate cleavage*: this occurs when separated blastomeres can each develop into complete animals, and is characteristic of most deuterostomes. It is regulatory development.
Gastrulation	Rearrangement of the blastula results in a gastrula with three germ layers: the endoderm, mesoderm and ectoderm. An archenteron cavity develops which will become the gut. In chordates the notochord differentiates from cells in the mid-dorsal region of the archenteron.
Neurulation and organogenesis	*Neural tube formation*: interactions between the notochord and ectoderm cells induce some ectoderm cells to form a neural tube which, in vertebrates, differentiates to form the brain and spinal cord. *Mesenchyme*: mesenchyme cells from the neural crest and mesoderm cells migrate and develop into sensory neurons, Schwann cells, adrenal medulla components, head muscles and skeleton, and teeth and scales. *Ectoderm*: placodes of the ectoderm develop into sensory organs, being induced to do so by underlying regions of the nervous system. *Mesoderm*: the mesoderm gives rise to the dermis, most of the skeleton and body muscles, the kidneys and the circulatory system. *Endoderm*: the endoderm gives rise to the gut, secretory cells in its associated organs, and the lungs.
Sex determination	Sex is determined by the presence of sex chromosomes with relevant genes, by ratios of sex to autosomal chromsomes, or by environmental cues such as egg incubation temperature.
Development modes	In indirect development a larval stage is present; direct development occurs in an egg envelope and a miniature adult (juvenile) hatches.

Brooding	In oviparous animals the eggs develop outside the mother's body, with or without parental care. Viviparous animals retain the young within the parent until development is completed. A placenta to supply nutrients may be present, with varying degrees of sophistication.
Extra-embryonic membranes	*Cleidoic eggs*: cledoic (self-contained) eggs, as in birds, have yolk, albumen and a shell. *Extra-embryonic membranes*: four extra-embryonic membranes form the yolk-sac, amnion, chorion and allantois. *Placental mammals*: in placental mammals (and marsupials), the outer trophoblast cells divide to form a syncytiotrophoblast which produces enzymes to facilitate implantation in the uterus wall. Cavities form in the inner cell mass above and below the embryonic disk to serve as an amnion and a yolk-sac. The trophoblast is the equivalent of the chorionic ectoderm, later augmented by mesoderm cells. The allantois extends to carry the fetal blood vessels to the chorion. The chorionic exchange area increases by the formation of chorionic villi which become the exchange surfaces of the placenta. In humans, the number of tissue layers separating the maternal blood and fetal capillaries decreases and the uterus wall breaks down so that maternal blood enters lacunae developing within the syncytiotrophoblast Placental histology varies between mammalian species.
Birth	Uterine contractions expel the mammalian fetus and placenta and facilitate reduction in the size of the uterus. Triggering is probably by hormonal changes.
Related topics	Body plans and body cavities (B1) Reproduction, gametogenesis and Protostomes and deuterostomes (B4) fertilization (D1) Integration and control: hormones (C22)

Egg types

Yolk enables the embryo to survive until it can feed by itself after hatching or until a placenta has been formed. Eggs can be categorized as:

- **isolecithal**: a small amount of yolk evenly distributed in the cytoplasm, for example mammals, echinoderms;
- **telolecithal**: yolk is concentrated at the vegetal (lower) pole; the animal (upper) pole contains the nucleus and less yolky cytoplasm, for example flatworms, most molluscs, polychaete annelids, most vertebrates. In some cephalopods, fishes, reptiles and birds, the yolk may comprise 90% of the egg, the nonyolky cytoplasm and the nucleus being confined to a cap at the animal pole;
- **centrolecithal**: yolk is concentrated in the egg center, with nonyolky cytoplasm outside and in the very center (containing the nucleus), for example arthropods.

Cleavage patterns

After fertilization, the zygote undergoes **cleavage** to form a **morula** (a cluster of cells, **blastomeres**) and then a **blastula**, a hollow ball of cells surrounding a **blastocoel**. In isolecithal and moderately yolked telolecithal eggs, the whole zygote cleaves: this is **holoblastic cleavage** (e.g. amphioxus and frog). In heavily yolked telolecithal eggs and in centrolecithal eggs, cleavage is incomplete or **meroblastic**.

Holoblastic cleavage in the **isolecithal egg** of amphioxus is **radial**, with the initial three cleavage planes mutually at right angles to each other, to give eight blastomeres of roughly equal size. The embryo can be bisected along any vertical (meridional) plane to give two mirror image halves; blastomeres below the equator may be larger in some species if more yolk is concentrated near the vegetal pole. The blastocoel is situated almost centrally, with a single wall of cells.

In the moderately telolecithal frog egg, the first two cleavage planes are meridional. At fertilization, pigment in the animal hemisphere opposite the point of entry of the sperm is withdrawn to leave a **gray crescent**: this identifies the posterior end of the frog. The first cleavage plane bisects the gray crescent and delineates the right and left halves of the animal. The larger amount of yolk in the vegetal hemisphere results in the third, equatorial cleavage plane being nearer the animal pole. The four upper, smaller blastomeres are termed micromeres, the four lower, larger blastomeres are macromeres. The blastocoel is displaced upwards towards the animal pole and its walls comprise several layers of cells.

In mammals, the eggs are secondarily isolecithal and cleave completely, albeit more slowly and irregularly; the blastula is known as a **blastocyst,** comprising an outer epithelium of **trophoblast** (which will develop into the placenta) surrounding the fluid-filled blastocoel cavity and a group of blastomeres at one pole that constitutes the **inner cell mass**.

In annelids and in molluscs, excluding cephalopods, holoblastic cleavage is **spiral**. The first and second cell divisions are meridional to form four macromeres. The third division has an oblique cleavage plane, resulting in four upper micromeres displaced circularly. The net effect is that when the egg is viewed from above the cleavage pattern appears as a spiral.

In the heavily yolked **telolecithal egg** (e.g. a bird's egg), only the superficial disk of cytoplasm at the animal pole undergoes meroblastic, **discoidal cleavage**, initially to form a single-layer **blastoderm**, followed by equatorial cleavage. In reptiles and birds a sub-germinal cavity appears beneath the blastoderm into which blastoderm cells migrate to form a hypoblast layer, leaving an epiblast with a blastocoel cavity between the epiblast and the hypoblast (*Fig. 1*).

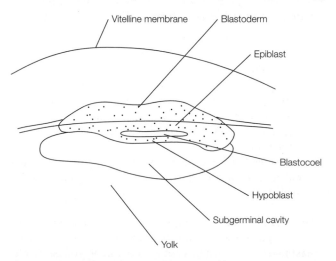

Fig. 1.   Discoidal cleavage in a bird egg.

In **centrolecithal eggs** (e.g. those of insects), the zygote nucleus lies in the center of the egg in an island of cytoplasm. After several nuclear divisions, the daughter nuclei (with some cytoplasm) migrate out to the peripheral cytoplasm where the central and peripheral cytoplasm fuses. The nuclei are present in a syncytium (i.e. no cell membranes between them). Eventually the cytoplasm divides to form a cellular blastoderm surrounding the yolk: this is the blastula. The insect cleavage is known as **superficial cleavage**.

Radial cleavage (whether holoblastic or meroblastic) of the fertilized egg is characteristic of **deuterostome** phyla such as Echinodermata and Chordata. **Spiral cleavage** (whether holoblastic or meroblastic) is characteristic of **protostome** phyla such as Annelida, Arthropoda and Mollusca (see Topic B4).

Mosaic and regulatory control

Determinate cleavage
This is found in urochordates such as sea-squirts. The prospective fates of blastomeres following early cleavage of the zygote are rigidly determined. For example, if one blastomere is removed from the embryo very early in development (e.g. eight cells), this may result in the subsequent absence of a notochord. That each blastomere goes on to form a specific part of the embryo is termed **mosaic** development. Such determinate cleavage is also characteristic of animals which show spiral cleavage, such as arthropods and molluscs.

Indeterminate cleavage
This is seen in the sea-urchin (an echinoderm). If early blastomeres of sea-urchins are separated, each will develop into a complete larva. Thus the variety of tissues into which each blastomere can potentially develop is much more diverse than would normally occur. Such development is described as **regulative** and is characteristic of echinoderms and most chordates (excluding urochordates).

Gastrulation

After the blastula has formed, **gastrulation** takes place: the cells of the blastula are rearranged following cellular movements to bring cells into positions appropriate for their future development. Radial symmetry is superseded by bilateral symmetry (see Topic B2). The blastula is succeeded by a **gastrula** (*Fig. 2*).

The gastrula has three layers of epithelium: the three **germ layers**:

(1) **endoderm** forms the gut lining;
(2) **ectoderm**, on the outside, forms the epidermis and nervous system;
(3) **mesoderm**, between the ectoderm and the endoderm (in triploblastic animals), forms most of the muscles and skeleton, and the circulatory and excretory tissues.

In isolecithal eggs (e.g. the deuterostome amphioxus), cells at the vegetal pole invaginate (push in) and an **archenteron** cavity is formed; the archenteron will become the gut. The opening into the archenteron, at the site of invagination, is the **blastopore**. In deuterostomes such as echinoderms and chordates, the blastopore becomes the anus. The blastocoel is subsequently obliterated. Initially, the outer layer of the gastrula is ectoderm, the inner layer, lining the archenteron, is endoderm; a strip of cells in the archenteron roof becomes the mesoderm (*Fig. 2*), although the derivation of mesoderm is very variable in different species.

The gastrula elongates and, in amphioxus, the notochord differentiates as a longitudinal rod of cells in the mid-dorsal region of the archenteron. The

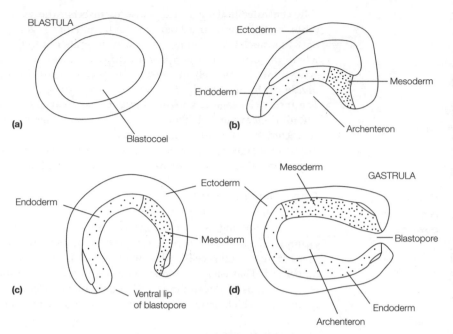

Fig. 2.   Longitudinal sections of amphioxus embryo to show gastrulation.

mesoderm forms as a series of pouches from the archenteron roof, on either side of the notochord (*Fig. 3*), losing the connection with the gut. The pouches fuse and their cavities fuse in amphioxus to form an **enterocoelic coelom** (see Topic B1).

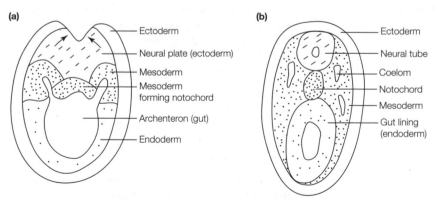

Fig. 3.   Development of tissues in amphioxus.

In amphibians (with moderately telolecithal eggs), cells at the margin of the gray crescent in the egg invaginate, those above the invaginating cells forming the dorsal lip of the blastopore. The cells of the animal hemisphere then roll over the lip of the blastopore and **involute** inwards to deepen the archenteron (*Fig. 4*): the blastocoel is gradually obliterated. Endoderm forms from the yolky cells in the vegetal hemisphere: these form the floor of the archenteron. The lip

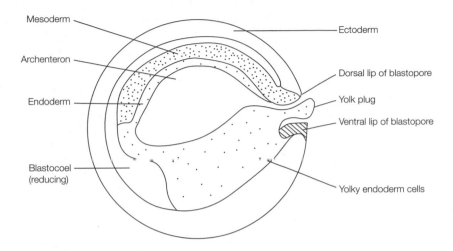

Fig. 4. Frog gastrula.

of the blastopore covers the yolk-laden cells until a blastopore is formed surrounding a yolk plug of yolk-laden cells. The blastopore becomes the anus, or it closes and the anus develops from a new invagination adjacent to the closed blastopore, according to the species of deuterostome.

In the heavily telolecithal bird egg, the germ layers develop from the epiblast (or embryonic disk, in mammals). Cells are carried towards the mid-dorsal, longitudinal **primitive streak**; they move inwards and spread out laterally into the blastocoel. There is no blastopore as such, the cells **ingressing** through the primitive streak which is probably the blastopore homolog. The first ingressing cells become endoderm. These replace the hypoblast and spread over the mass of yolk. Later, ingressing cells form the mesoderm and the notochord, while those remaining on the epiblast surface become the ectoderm.

In protostome animals (e.g. molluscs and arthropods), the blastopore gives rise to the mouth (see Topic B4); the mesoderm develops from cells which arose before gastrulation and which then move (ingress) into the blastocoel to lie alongside the archenteron, forming mesodermal sheets between the ectoderm and mesoderm: the **schizocoelic coelom** forms by the splitting of these sheets (see Topic B1).

Neurulation and organogenesis

Neural tube formation
The **notochord** differentiates longitudinally from mesodermal cells in, or adjacent to, the archenteron roof. Interactions between the mesoderm forming the notochord and overlying ectoderm cells **induce** ectoderm cells to form a **neural tube**. The ectoderm cells first thicken to form a longitudinal neural plate, in the center of which a neural groove develops. The outer edges of the groove rise up as folds. The dorsal parts of the fold meet, enclosing (from the anterior end backwards) a neural tube (*Fig. 5*). In craniates (vertebrates), the anterior part of the neural tube expands to form the brain. The enclosed cavity becomes, anteriorly, the ventricles of the brain, and posteriorly, the neural canal of the spinal cord. The deeper parts of the folds form the neural crest. This process is termed **neurulation**, as the gastrula develops into a **neurula**.

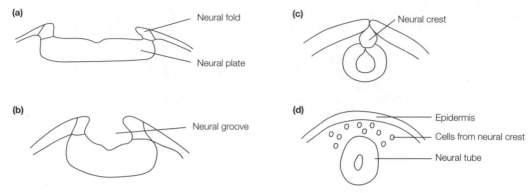

Fig. 5.   Cross-sections to show neural tube formation in the frog embryo.

Mesenchyme cells

Some neural crest and mesoderm cells become **mesenchyme** cells which migrate and differentiate into:

- sensory neurons
- postganglionic autonomic neurons
- Schwann cells
- components of the adrenal medulla
- head muscle and skeleton
- teeth and scales

Ectoderm

The olfactory regions of the nose, together with the eyes and ears, form from **placodes** of ectoderm: placode differentiation and development into the sensory organ are the result of inductive interactions with underlying regions of the nervous system.

Mesoderm

Mesoderm gives rise to most of the dermis, skeleton and muscles, the kidneys and circulatory system. Segmental **somites** develop sequentially, from the anterior to the posterior of the animal, alongside the notochord and neural tube. In craniates (vertebrates), mesenchymal cells from the most medial part of each somite migrate around the neural tube to form the vertebrae and cranium. The notochord regresses, remaining as the intervertebral discs in higher vertebrates. Dorsolateral somite cells form the dermis; the central part of the somite expands to form a muscle block or **myotome** which forms the major skeletal body muscles. Mesoderm ventral to the somites and next to the coelom forms a **nephric ridge**, giving rise to kidney tubules. The **lateral plates** of the somites, lying on each side of the coelom, give rise to parts of the lining of the coelom and to muscles of the flank and the gut.

Blood vessels originate in the mesoderm over the yolk as **primitive blood islands**: epithelial cells surround other cells which will become the first blood cells. The islands aggregate to form vessels. Paired vitelline veins fuse beneath the developing pharynx to form the **heart**, which beats with its own myogenic rhythm very early in organogenesis.

Endoderm

Endoderm cells are the origins of the lining to the gut and lungs, and the secretory cells of the pancreas and the liver. The lungs, liver and pancreas originate as pouches off the gut: the muscles and blood vessels of these organs derive from mesoderm. The anterior part of the archenteron develops into the pharynx: paired pharyngeal pouches develop laterally in the endoderm lining and meet corresponding invaginations of ectoderm. These form the **gill-slits**. The mouth and cloacal cavities are usually formed by inpocketings of ectoderm which meet the endoderm-lined archenteron.

Sex determination   Sex is determined by one of the following:

- the presence of sex chromosomes;
- the ratio of sex chromosomes to autosomal chromosomes;
- environmental cues.

Mammals possess **X and Y sex chromosomes**: males are XY and females are XX. [In reptiles and birds the male is WW and the female WZ, W and Z designating the sex chromosomes.]

In mammals, the early embryo has primordial gonads, gonoducts and genitalia which are **morphologically indifferent** (i.e. they could develop into either male or female reproductive systems). The Y chromosome carries a gene (**TDF** – testicular-determining factor gene – in humans) which directs the embryonic gonad to develop as a testis. (Thus an XXY individual is morphologically a male and an XO individual is morphologically a female.) Subsequently, further differentiation is controlled by hormones: Leydig cells in the fetal testes secrete testosterone, and an anti-Müllerian duct factor, **AMDF**, is produced by Sertoli cells in the testes, facilitating the regression of the fetal oviduct (Müllerian duct). The *absence* of TDF and AMDF promotes the development of a female reproductive system.

In the fruitfly *Drosophila* sp. it is the **ratio of X chromosomes to autosomal chromosomes** which determines sex. Furthermore, differentiation is directed by direct gene action rather than by hormones. Other Diptera (flies) use other mechanisms: in the housefly there seems to be a male-dominant-determining factor on the Y chromosome. In honeybees, if the egg is fertilized it becomes a female, and if it is not fertilized it becomes a haploid male

In some species, **environmental cues** determine sex. **Position** affects sex determination in echiuran worms, a female developing if the larva settles on the substratum without contacting another worm; if contact does occur between a female and one which has already settled, the latter secretes a substance which makes the larva develop into a minute male which lives within the female's reproductive tract. **Egg incubation temperature** affects sex in some reptiles. In *Alligator mississippiensis* lower temperatures ($\leqslant 30$ºC) favor females and higher temperatures ($\geqslant 34$ ºC) males. Incubation at 32ºC produces an 87:13 female:male ratio.

Development modes   The presence of a **larva** allows dispersal of the young, particularly important in sessile species.

- **Indirect development**: a larval stage is present.
- **Direct development**: the larval stage is dispensed with; development occurs within an egg envelope, and miniature adults (juveniles) hatch.

Brooding

In **oviparous** animals eggs develop outside the mother's body. Some parents care for their eggs by **brooding**, as in birds.

In **viviparous** animals the young leave the parent when development is completed. A few animals brood their eggs internally in a modified region of the oviduct (e.g. some salamanders and snakes): in the seahorse the male has an abdominal brood pouch. **Aplacental** or **ovoviviparous** animals (e.g. many sharks) have a viviparous development where all the nutrient requirements of the young are contained in the egg. There is a continuum to complete **placental viviparity** (e.g. scorpions and therian mammals) in which there is transfer of nutrients across a **placenta**.

Extra-embryonic membranes

Cleidoic eggs

In most terrestrial animals (e.g. birds, reptiles and insects), the egg is **cleidoic** (or 'self-contained'): within it is food including fats which can be metabolized to provide water. Waste is stored within the egg membranes. The egg is bounded by a waterproof but gas-permeable shell. Reptile, bird and monotreme mammal eggs have yolk surrounded by albumen and a shell, secreted by the oviduct.

Extra-embryonic membranes

Four extra-embryonic membranes have evolved (*Fig. 6*):

- the **yolk-sac** surrounds the yolk;
- the **amnion** encloses the embryo and its amniotic fluid-filled cavity bathes the embryo in a watery environment;
- the **chorion** lies beneath the shell and encloses the yolk-sac and the amnion;
- the **allantois** grows out from the posterior end of the gut and lies against the inner side of the chorion; blood circulates through vessels in the chorionic wall and here gas exchange occurs. The allantois stores waste uric acid.

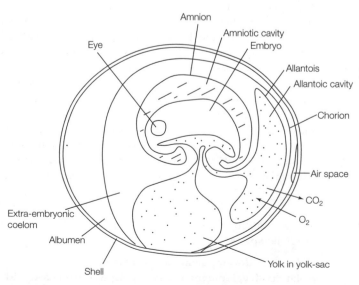

Fig. 6.   Extra-embryonic membranes in a bird egg.

All membranes except the chorion connect to the center of the ventral region of the embryo; the yolk-sac is resorbed during embryogenesis.

Placental mammals
In the **placental mammals**, fertilization occurs in the upper part of the oviduct, and the **conceptus** cleaves as it passes down the duct. Outer trophoblast cells mutiply without cytoplasmic division to form a **syncytiotrophoblast**, the inner trophoblast remaining cellular. The syncytiotrophoblast produces enzymes allowing the embryo to digest the uterine lining and **implant**. Cavities form in the inner cell mass above and below the **embryonic disk** (the prospective embryo). The upper cavity is the amnion, the lower is the yolk-sac. The trophoblast is the homolog of the chorionic ectoderm in other vertebrates. During gastrulation, mesoderm cells move around the inside of the trophoblast to complete the chorion, and around the amnion and the yolk-sac (which is derived from endoderm). The allantois finally extends to carry fetal blood vessels to the chorion (*Fig. 7*). The exchange area of the chorion is extended by **chorionic villi** and, in many species, including humans, the number of tissue layers separating the maternal and fetal circulations reduces.

In humans, the syncytiotrophoblast, on the chorionic side of the conceptus, continues to invade and break down the uterus wall including the uterine blood vessels. Maternal blood enters **lacunae** developing within the syncytiotrophoblast; chorionic villi with fetal capillaries in them grow into the lacunae and in this **placenta** facilitate exchange of gases, nutrients, waste and other substances (e.g. hormones, IgG antibodies, alcohol). The placental barrier comprises, in humans, the endothelial walls of the fetal capillaries, trophoblast and connective tissue. The placenta forms an immunological barrier between

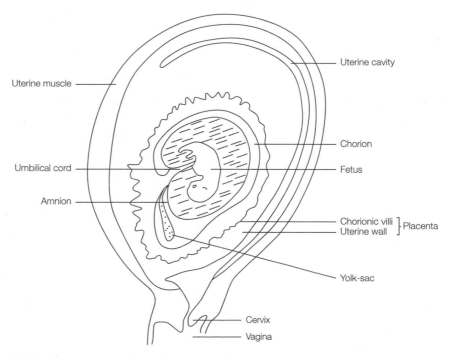

Fig. 7. The mammalian fetus in the uterus.

mother and fetus and secretes hormones (e.g. chorionic gonadotropin and progesterone) which help to maintain pregnancy.

Birth

Involuntary contractions of uterine smooth muscle initiate **labor**; the fetus in its amniotic sac is pushed against the neck of the cervix which dilates. The amniotic sac ruptures to release amniotic fluid ('breaking of the waters') and the fetal head is pushed into the vagina, stimulating stronger uterine contractions by a process of positive feedback. Birth follows. Uterine contractions expel blood from the placenta to the **neonate** via the umbilical cord. The placenta is then expelled. The umbilical vessels contract and the mammal then bites through the cord (in humans the cord is cut!). Uterine contractions after birth reduce the size of the uterus and help to prevent excessive bleeding.

The triggering of the birth process is not fully understood, but in sheep it seems to be linked to endocrine changes in the fetal lamb: fetal cortisol affects secretion of placental estrogens, and activity is initiated in the myometrium of the uterus wall, probably by local increases in prostaglandin levels.

D3 LACTATION

Key Notes

Lactation	Lactation comprises lactogenesis (initiation of milk secretion) and galactopoiesis (maintenance of lactation).
Mammary glands	Mammary glands enlarge during pregnancy and alveoli develop under the influence of hormones.
Lactogenesis	Prolactin stimulates lactogenesis: during pregnancy placental estrogens and progesterone inhibit milk production; the levels of these hormones are reduced following birth.
Galactopoiesis	Galactopoiesis is under the control of several hormones. Birth stimulates colostrum release, and this is replaced by milk after 24–72 hours. Colostrum is rich in proteins, hormones and laxatives. Colostral (and milk) IgA levels are high in humans, which confers passive immunity to the baby to help combat gut pathogens.
Suckling	Suckling stimulates the hypothalamus, leading to a prolactin surge which stimulates further milk production, and oxytocin release which stimulates milk expulsion.
Related topic	Hormones (C22)

Lactation   Lactation (milk production in mammals) comprises two phases: **lactogenesis** (the initiation of milk secretion) and **galactopoiesis** (the maintenance of established lactation).

Mammary glands   Ducts in the mammary glands develop after puberty, under the influence of estrogens, growth hormone and adrenal glucocorticoids. During pregnancy, the mammary glands enlarge and **alveoli** develop in response to increasing levels of estrogens, growth hormone, adrenal glucocorticoids, insulin, progesterone and (from the placenta) somatomammotropin.

Lactogenesis   Prolactin (from the anterior pituitary) stimulates lactogenesis; before birth placental secretions of estrogens and progesterone inhibit milk production. After the expulsion of the placenta from the womb, estrogen and progesterone levels fall and lactogenesis commences.

Galactopoiesis   Continued galactopoiesis requires growth hormone, parathyroid hormone and cortisol which facilitate provision of amino acids, fatty acids and calcium.

In women, milk release does not begin until 24–72 hours after birth, the expulsion of the baby being the stimulation. During the first 24–72 hours the baby receives **colostrum**. Colostrum has high levels of proteins and is present in the mammary glands when the baby is born. Colostral components include laxatives which encourage the elimination of fetal wastes (e.g. mucus, bile and dead gut epithelium cells) from the gut. Hormones to facilitate development of the neonatal digestive system are contained in colostrum (and, later, milk).

Human colostrum (and, later, milk) has high levels of IgA antibodies and confers passive immunity to the infant. The maternal IgA remains in the gut where it neutralizes gut pathogens: IgA, greatly helped by other colostral components such as lactotransferrin, acts on harmful microorganisms such as coliform bacteria (e.g. *Escherichia coli*): breast-fed human babies have less than 20% of the mortality of bottle-fed babies from coliform pathogens. Immune cells (macrophages, neutrophils and lymphocytes) are also transferred in colostrum and milk. Such B lymphocytes include IgA-secreting cells, particularly responsive to gut pathogens (e.g. *E. coli* K1 strain which causes 84% of infant meningitis). Other species may not transfer IgA; cow milk contains maternal IgG which can be transported across the calf gut wall to enter the calf's circulation.

Human milk comprises 88% water, 6.5–8% carbohydrate, 1–2% protein, 3–5% fat and 0.2% salts. Milk in marine mammals (e.g. seals and whales) is very rich in fats and enables the newborn animals to build up insulating blubber very quickly. Seal milk contains 57.9 g of fat per 100 g of milk, compared with 3.7 g of fat per 100 g of cow's milk.

Suckling

Nervous signals from the sucking action on the nipples or teats pass to the mother's hypothalamus and cause a surge of **prolactin** from the anterior pituitary gland which lasts, in women, for about 60 minutes. This acts on the mammary glands to provide milk for the next suckling period. Stimulation of the hypothalamus by sucking also results in the secretion of **oxytocin** via the posterior pituitary. Oxytocin causes the smooth muscles in the mammary glands to contract, expelling milk from the alveoli: stronger sucking results in stronger alveolar myoepithelial cell contraction and milk expulsion, an example of positive feedback. (Mammary glands which are not emptied soon stop galactopoiesis: in women, lactation can continue for over 2 years, although the rate of galactopoiesis slows after about 8 months.) See *Fig. 1*.

Lactation helps the uterus to contract following birth by suppressing production of gonadotropins and sex steroids; menstruation may not occur in nursing mothers, although over 50% of women do ovulate while lactating.

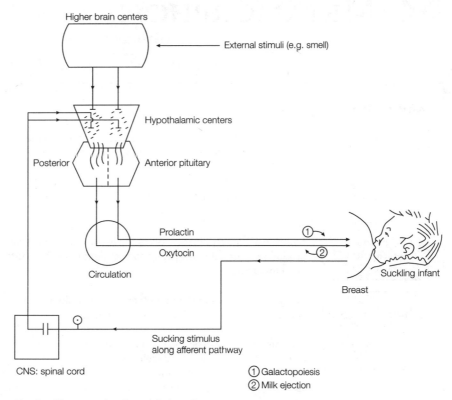

Fig. 1.   *The control pathways in lactation.*

D4 METAMORPHOSIS

Key Notes

Metamorphosis	Metamorphosis is a radical transformation from one life form to another during an animal's life cycle, usually associated with habitat and life style changes.
Insect metamorphosis	Insect metamorphosis is promoted by ecdysone in conjunction with juvenile hormone.
Tunicate (urochordate) metamorphosis	Tunicate metamorphosis involves the loss of the typical chordate structure to produce a (normally) sessile individual dominated by a filter-feeding branchial basket.
Teleost metamorphosis	Many teleost fishes undergo metamorphosis linked with changes in habitat, apparently associated with endocrine changes.
Amphibian metamorphosis	Amphibian metamorphosis is under the control of the hypothalamus–pituitary–thyroid axis. The development of hypothalamus–pituitary portal capillaries allow prolactin-inhibiting factor to inhibit prolactin release, and allow thyrotropin-releasing factor to stimulate thyroid-stimulating hormone production, leading to thyroid hormone release, promoting metamorphosis.
Neoteny in urodeles	Many urodeles are facultative neotenes, with sexually mature larvae which only metamorphose under certain conditions; others are perennibranchiates which are never observed to metamorphose.
Related topics	Neoteny and pedogenesis (B6) Hormones (C22) Evolutionary origins of the Chordata (B7)

Metamorphosis

Metamorphosis is the transformation from one to another morphological and physiological form of an animal during its life cycle; true metamorphosis is concerned with preparation for life in a different habitat (e.g. fresh water as opposed to sea, land as opposed to water). Changes may be associated with migration and/or sexual maturity.

Insect metamorphosis

In holometabolous insects (e.g. butterflies), the immature larva (caterpillar) is very different from the adult insect in body form, habitat and behavior. There are a fixed number of species-specific larval molts, the last resulting in a pupa (often contained within a protective case or cocoon of silk) in which radical changes in body organization take place. (The chrysalis is the larval exoskeleton retained in some forms through the pupal stage.) The final molt occurs within the cocoon and the adult (imago) emerges.

During larval life, the larval condition is maintained by a juvenile hormone secreted by the corpus allatum. Environmental cues such as light or temperature stimulate the brain to synthesize a brain hormone which acts on the prothoracic gland to release ecdysone which stimulates pupation and metamorphosis.

Tunicate (urochordate) metamorphosis

The urochordate larva has a typical chordate structure with a notochord, pharyngeal gill-slits, dorsal nerve cord and segmental muscles in a post-anal tail. Following metamorphosis, most of these features are lost, the tunicate settles to become sessile, the pharnyx expands to form a complex branchial basket for filter-feeding, and an ovotestis develops. [A speculation that during evolution a tunicate larva may have become sexually mature through a process of pedogenesis (see Topic B6) to give rise to an ancestral vertebrate is important in theories of chordate phylogenesis (see Topic B7).]

Teleost metamorphosis

Eels and salmonids can be said to undergo metamorphosis: the leptocephalus eel larva transforms into an elver, but little is known of marine transformations prior to a return to rivers. Salmon spawn in fresh water and the young parr, on reaching a certain size, transform into smolt which migrate to the sea. Endocrine changes are seen in the histology of the pituitary and thyroid glands. Levels of 17-hydroxycorticosteroids in parr are five times those in smolt, and iodine uptake into the thyroid increases early in metamorphosis.

Amphibian metamorphosis

Anura
The metamorphosis from the larval tadpole to the adult toad or frog involves many profound physiological and gene-regulatory changes. Larval life can vary in time from a few weeks in the spadefoot toad to several years in the bullfrog. The time of metamorphosis is affected by season and geography: Florida bullfrogs (*Rana catesbeiana*) metamorphose during their first summer, northern bullfrogs take 3 years. Nutrition, crowding and water temperature and pH are known to affect timing.

The role of the endocrine system in amphibian metamorphosis is deduced from the fact that thyroxine (T_4), tri-iodothyronine and TSH can initate metamorphosis; antithyroid drugs (e.g. propylthiourea) prolong larval life or prevent metamorphosis.

During larval life, levels of T_4 are low and levels of PRL are high. As larval life proceeds, portal capillaries develop between the hypothalamus and the anterior pituitary. TRF from the hypothalamus can now reach the anterior pituitary which synthesizes TSH, in turn stimulating the thyroid to manufacture and release T_4, stimulating metamorphosis. (Hypophysectomized tadpoles in which the pituitary has been removed do not metamorphose.) There is some doubt about the role of TRF in amphibians, but if a tadpole pituitary is transplanted into the tadpole tail metamorphosis does not occur, suggesting that the hypothalamus–pituitary portal link is important.

Injection of PRL antagonizes T_4-induced tail resorption and delays metamorphosis: thus high levels of PRL in tadpoles probably exert an inhibiting effect on the thyroid. Development of the hypothalamus–pituitary link also allows prolactin-inhibiting factor (PIF, which may be dopamine) probably made in the hypothalamus, to reach the pituitary.

Sequence

- *Pre-metamorphosis*: rapid growth occurs. PRL levels are high. Thyroid hormones are low; the hypothalamus–pituitary link is poorly developed. Therefore, there are low TSH and PIF levels.
- *Pro-metamorphosis:* the hypothalamus–pituitary link develops. TRF (?) and PIF reach the anterior pituitary. TSH and T_4 levels rise, and PRL levels fall. The hypothalamus continues to mature under positive feedback of T_4. Massive stimulation of the thyroid leads to metamorphic climax.
- *Post-metamorphosis*: T_4 levels fall under negative feedback to adult levels.

Urodela

Urodeles (newts, salamanders, etc.) show a less dramatic metamorphosis. Loss of external gills is the main feature, but tail resorption and head shape changes are also seen. As in anurans, urodelan metamorphosis is under the control of the hypothalamus–pituitary–thyroid axis and can be inhibited by PRL.

The eastern spotted newt (*Notophthalmus viridescens*) shows a **double metamorphosis**, the first from an aquatic larva to a terrestrial 'red eft', the second when it returns to water after 2–3 years: there are changes in the skin and the tongue is lost. PRL levels rise and T_4 levels fall after the return to water.

Neoteny in urodeles

Neoteny is common in urodeles. Axolotls (*Ambystoma mexicanum*) are **facultative neotenes**. They metamorphose under drought conditions or if the hypothalamus–pituitary–thyroid axis is stimulated, for example with TSH: the larval form can be sexually mature (see Topic B6). Pituitaries from tiger salamanders (an **obligate metamorphoser**) will cause axolotls to metamorphose if transplanted into them. Neotenic axolotl pituitaries will not cause hypophysectomized tiger salamander larvae to metamorphose. This suggests activity of the hypothalamus–pituitary–thyroid axis is too low to permit metamorphosis in facultative neotenes. (It is interesting that TRF *is* found in high levels in axolotl hypothalami, but seems to be ineffective in stimulating TSH and thereby metamorphosis.)

Perennibranchiate urodeles can never be induced to metamorphose, however much T_4 is given (e.g. in the ditch-snake, *Amphiuma means*). Thyroid extracts from the perrenibranchiate *Necturus* sp. will facilitate metamorphosis in *Salamandra* sp. Receptors to pituitary–thyroid products are presumably lost in perennibranchiates.

A **cryptometamorphosis** does occur in neotenic axolotls: there is a change from the larval to the adult hemoglobin which is not connected to morphological metamorphosis (a similar phenomenon is noted in the anuran *Xenopus laevis* when tadpole metamorphosis is blocked with propyltiourea): this change may be unrelated to hypothalamus–pituitary–thyroid axial activity, or be driven by very low levels of such activity.

D5 PUBERTY IN HUMANS

Key Notes

Puberty onset

Puberty is characterized by menarche in girls and the first emission of semen in boys.

Growth and body dimensions

Following a minimal growth rate in late childhood, there is an adolescent growth spurt (about 2 years later in boys than girls), succeeded by decreased growth and epiphyseal fusion in the bones. Sexual dimorphism is seen in muscle, bone dimensions and fat deposition.

Secondary sexual characteristics and sexual maturity

Puberty is associated with the development of secondary sexual characteristics and sexual maturity, including gametogenesis.

Hormonal control of puberty

Control is hormonal through gonadotropin-releasing hormone from the hypothalamus stimulating gonadotropin (luteinizing hormone and follicle-stimulating hormone) synthesis and release from the anterior pituitary. These stimulate the synthesis and release of sex steroids from gonads. In childhood, the portal link between the hypothalamus and the pituitary is undeveloped, so the hypothalamus–pituitary–gonad link is incomplete.

Adrenarche

Adrenal androgens synthesis commences in later childhood (the adrenarche).

Trigger for puberty onset

The trigger for the timing of puberty is probably body weight.

Related topics

Hormones (C22)
Metamorphosis (D4)

Menstrual cycle in women (D6)
Mid-life and menopause in humans (D7)

Puberty onset

Puberty is a period of rapid body growth and development. accompanied by the secretion of gonadal hormones and the development of sexual maturity, including the production of gametes and the development of secondary sexual characteristics.

In girls, the onset of puberty is marked by the **menarche**, the first menstrual (monthly) period: the ovary has secreted sufficient steroid hormones to induce uterine development, although ovulation may not occur until later. Menarche is preceded by a series of changes over 2–5 years, dependent on secretion of sex steroids from the ovaries and the adrenals. The sequence of changes is constant, but can take varying periods of time and start at different ages.

In boys, the onset of puberty is harder to define, although the first emission of semen could be taken as a starting point: there are probably no viable spermatozoa yet.

Growth and body dimensions

- There is minimum growth velocity in a child before **'take-off'** up to the **adolescent growth spurt**.
- There is maximum growth velocity (especially height) during the adolescent growth spurt
- This is followed by decreased growth velocity; cessation of growth following **epiphyseal fusion** (consolidation and fusion at the centers of bone growth)

Boys start their growth spurt about 24 months after girls: they are therefore taller at 'take-off' (the start of the adolescent growth spurt). Height gain during the adolescent growth spurt is similar in girls (25 cm) and boys (28 cm). Thus the mean 10 cm height difference between men and women is mainly due to height differences before the adolescent growth spurt starts.

Virtually all muscle and skeletal dimensions are involved in the adolescent growth spurt, but there is extensive sexual dimorphism in growth rates of different body regions (e.g. shoulders in boys and hips in girls); fat deposition shows different patterns in boys and girls. Lean body mass and body fat are approximately the same in boys and girls before puberty; men have about 1.5 times the lean body mass and 1.5 times the skeletal mass of women (a man's greater strength is due to his possession of more muscle cells); women have about twice the body fat of men; these changes start at about 7 years in girls and about 9 years in boys.

Secondary sexual characteristics and sexual maturity

Puberty enhances secondary sexual characteristics (breast development in girls, penis and scrotum enlargement in boys, thickening of vocal cords in boys, growth of facial hair in boys and body hair in boys and girls). Gametogenesis is initiated in the final stages of puberty.

Hormonal control of puberty

Control is by the **hypothalamus–anterior pituitary–gonad axis** (*Fig. 1*). In childhood, the **portal capillary** link between the hypothalamus and the anterior pituitary is undeveloped. Development of this link in late childhood allows gonadotropin-releasing hormone (**GnRH**) to reach the pituitary which synthesizes and releases **gonadotropins** (**LH** and **FSH**). These act on the gonads to stimulate synthesis and release of **sex steroids** (e.g. estradiol, testosterone). Estradiol enhances pituitary PRL release in girls.

An alternative therory states that increased body mass dilutes concentrations of melatonin from the pineal gland. Melatonin inhibits GnRH pulses in childhood. Pineal tumors with high levels of melatonin do delay puberty.

Pulsing GnRH into an immature female monkey leads to establishment of ovulatory cycles and 'puberty': the pituitary and gonads respond to the GnRH instantly and maintain a 'conversation' through feedbacks to the component parts of the hypothalamus–pituitary–gonad axis. If the GnRH is switched off, the monkey reverts to pre-puberty.

Thus levels of LH, FSH and PRL are low in childhood but during puberty they rise to adult levels. In early puberty, LH pulses during sleep; later it pulses during the day too. Such rhythms are not present in childhood but do occur in adulthood. Testosterone levels in the plasma are less than 10 ng 100 ml^{-1} in boys and girls before puberty; in early puberty, testosterone levels rise to 20–240 ng 100 ml^{-1} during sleep in boys. In girls, estradiol levels rise through puberty to reach 50 pg 100 ml^{-1} in the follicular stage of the menstrual cycle and 150 pg 100 ml^{-1} in the luteal stage. (In pubertal boys, estradiol levels are higher than during pre-puberty, but much lower than in pubertal girls.)

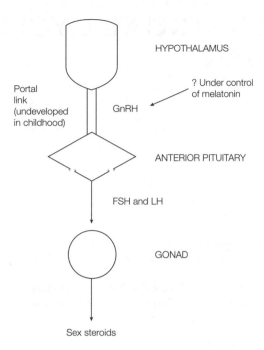

Fig. 1. The hormonal control of puberty: the hypothalamus–anterior pituitary–gonad axis.

Adrenarche

Adrenal androgens start to rise by 8 years of age, continuing to rise until 13–15 years in both girls and boys, preceding gonadotropin/gonadal steroid secretion. This **adrenarche** is one of the first pubertal hormonal events and promotes pubic and axilliary hair growth.

Trigger for puberty onset

Timing of puberty
Within individuals, the age of puberty varies enormously (e.g. in girls from 9 years to 16 years). The trigger for puberty is unknown. There has been a trend towards earlier menarche in girls and earlier puberty in boys (for girls in 1880 this occurred at 14–15 years, in 1980 at 12–13 years) although this may reflect biased sampling (e.g. deprived children in 19th century asylums). Higher social standards including **nutrition** and **better general health** have been suggested as causes for earlier menarche, as have **longer photoperiods** associated with the use of electric lighting (although blind girls attain the menarche no later than sighted girls).

   Nutrition seems a better candidate: indigenous peoples from northern Scandinavia who ate the same diet from 1870 to 1930 experienced no earlier puberty; patients with anorexia nervosa are amenorrhoeic (do not menstruate). Body weight at menarche is surprisingly constant at 47 kg in girls; the figure for boys is 55 kg. Thus body weight may trigger the onset of puberty (obese children have early puberty; gymnasts and ballet dancers with low weight to height ratios have delayed puberty). How such factors act on the development of the hypothalamus–pituitary links are not known, although increased body mass may 'dilute out' melatonin from the pineal gland (see above).

D6 THE MENSTRUAL CYCLE IN WOMEN

Key Notes

Pre-ovulation	The cycle begins with the shedding of the endometrial lining of the uterus. A subsequent increase in gonadotropins, FSH and LH, promotes the growth of the ovarian follicle and maturation of the egg.
Ovulation	Follicular estrogens promote regrowth of the endometrium and (by positive feedback) a surge of pituitary LH (and FSH) resulting in ovulation.
Post-ovulation	LH causes granulosa cells of empty follicles to enlarge to form a corpus luteum which secretes progesterone and also some estrogens which stimulate endometrial secretions (preparing the uterus for the implantation of the conceptus).
End of cycle	The absence of conception leads to corpus luteum degeneration and the shedding of the endometrium again.
Related topics	Hormones (C22) Mid-life and menopause in humans (D7)
	Puberty in humans (D5)

Pre-ovulation

The sequence of events is illustrated in *Fig. 1*. After puberty in women, the menstrual (monthly) cycle is established. The production of eggs (oocytes) is cyclical: it involves an interplay of hormones and changes in the follicle cells of the ovary and lining of the uterus. The controller is the **hypothalamus**. The hypothalamus secretes GnRH which acts on the anterior pituitary: this releases the gonadotropins LH and FSH. These act on the ovaries, stimulating them to make sex steroids, principally progesterone and estrogens. In low concentrations, **estrogens** act to increase the sensitivity of the pituitary to GnRH, resulting in increased gonadotropin synthesis; **progesterone** inhibits GnRH and gonadotropin production.

The **menstrual cycle begins** on the first day of **menstrual flow** (menorrhoea) caused by the shedding of the endometrium lining the uterus.

- Hormone levels are initially low.
- The increase in FSH and LH soon after the cycle starts promotes growth of the **ovarian follicle** and maturation of the egg.
- The follicle secretes estrogens which promote endometrial and follicular growth.

Ovulation

A rise in estrogens in mid-cycle due to dramatic growth of the follicles triggers a sharp increase in pituitary LH (positive feedback), stimulating release of the egg **(ovulation)**. FSH levels also rise, but less dramatically because they are regulated by another follicular hormone, **inhibin**.

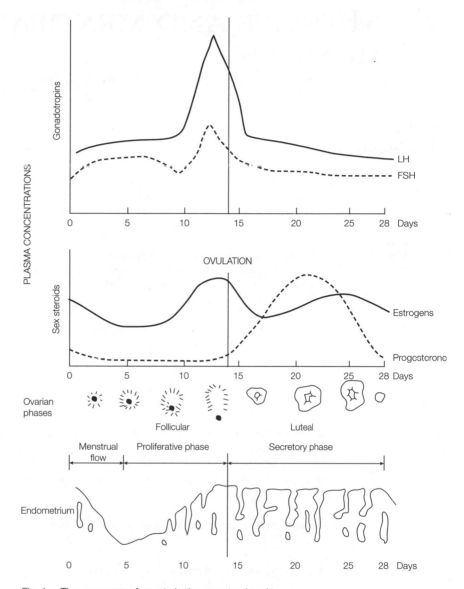

Fig. 1. The sequence of events in the menstrual cycle.

Post-ovulation

Under the continued influence of LH, cells from the emptied follicle enlarge and fill the cavity, producing the **corpus luteum** which secretes estrogens and progesterone. Progesterone further stimulates endometrial growth and transforms it into a secretory tissue, preparing it for implantation of the conceptus.

End of cycle

In the absence of fertilization of the egg, the corpus luteum degenerates, production of sex steroids falls, the endometrium is shed and the cycle ends and recommences.

During early and mid-adulthood, the woman's menstrual cycle continues with a periodicity of about 28 days.

D7 MID-LIFE AND MENOPAUSE IN HUMANS

Key Notes

Mid-life	Post-pubertal changes in body structure and function occur, including muscle weight loss, deposition of fat and loss of elasticity in body elements (e.g. collagen). Autoimmunity tends to increase and sensory acuity tends to decrease.
Menopause	Failure of the ovaries to function cyclically (climacteric) occurs at about age 52 in women, and cessation of menstrual flow indicates menopause. Hormonal changes (especially a fall in estrogens) affect a variety of physiological parameters and functions.
Related topics	Hormones (C22) The menstrual cycle in women (D6) Puberty in humans (D5)

Mid-life

Mid-life spans the years between puberty and old age (senescence). In men, maximum secretion of testosterone occurs between 18 and 20 years of age; muscular strength peaks at about 25 years of age. Changes are seen during different decades of life, although these **vary from person to person.**

Teens
Thymus involution occurs at about 12 years of age.

Twenties
Height may start to decrease due to a withdrawal of protein from the intervertebral disks.

Thirties
Women reach their sexual peak; heart muscles start to thicken; the skin loses some elasticity (collagen cross-linking) leading to wrinkles; melanin loss in hair leads to graying; further loss of protein from intervertebral disks occurs; loss of some hearing in the high-frequency range takes place.

Forties
Hunching of the back is seen: further loss of protein from intervertebral disks may mean that a 40-year-old man is about 4 mm shorter than he was at 20; there is a tendency to deposit fat (an average man is 4.5–9.0 kg heavier than he was in his twenties); there is a loss of scalp hair; and presbyopia is common.

Fifties

Menopause occurs in women (see below); late-onset diabetes and other auto-immune conditions are more common; the skin continues to loosen and wrinkle; hearing and chemoreception (taste and smell) decline; muscle deterioration and weight loss are common (although this may be more than compensated for by fat deposition as metabolism slows and less exercise is taken); semen volume decreases (although viable spermatozoa are produced into old age); men's voices may acquire a higher pitch due to stiffening of the vocal cords. These changes continue into the sixties and beyond.

Menopause

Women between 45 and 55 usually stop producing and releasing eggs and the monthly menstrual cycle (often after some irregularity of timing) ceases. This cessation of menstruation is the **menopause**, and failure of the ovaries to function cyclically is the **climacteric**: it signals the end of reproductive ability in women. In Western societies, the average age for the menopause is 52 years, and is increasing.

Ovarian failure results in low estrogens and hence raised levels of FSH and LH. The fall in estradiol production leads to atrophy of the breasts and the vaginal mucosae. Other changes include a decline in bone mass (sometimes leading to osteoporosis associated with reduced estradiol). 'Hot flushes' in the skin are caused by changes in the vasomotor system controlling cutaneous vasodilation. Dizziness, fatigue, headaches, chest and neck pains and insomnia are common. As estrogen levels fall (and blood cholesterol rises), the incidence of cardiovascular disease in women rises to equal that of men. Psychological disturbances (e.g. depression) may accompany the physical changes. After the menopause, estrogen production is solely dependent on conversions of adrenal steroids.

Hormone replacement therapy (HRT) can be used to ameliorate the effects of the menopause, although estrogen alone can result in endometrial hyperplasia and increase the risks of uterine cancers. Progesterone is given too: this reduces the numbers of estrogen receptors and increases the catabolism of estradiol. HRT is usually given intermittently and produces withdrawal bleeding from the uterus.

There is no evidence of a biologically based menopause-equivalent in men.

D8 AGING

Key Notes

Aging and sensescence	Aging (usually synonymous with senescence) implies a loss of vitality, a reduced ability to withstand environmental insults and an increased chance of dying with time.
Survivorship curves in populations	Population studies allow mortality distributions to be drawn on a graph on which the number of survivors is plotted against time. Hypothetical nonaging populations show a hyperbolic survivorship curve, whereas real aging populations show a rectangular suvivorship curve.
Aging in individuals	Aging in individuals is best measured using a battery of indices; complex physiological and homeostatic systems decline more than do simple systems.
Population studies	Point-in-time studies on a given date are easier, but samples tend to be self-selected for survival; longitudinal studies are logistically difficult.
In vivo studies	In vivo studies show that cells and tissues can be renewable from stem cells (e.g. gut lining cells), expanding (e.g. liver cells) or nonrenewable (e.g. brain neurons). Renewable cells can be serially transplanted from host to host and can outlive the donor.
In vitro studies	In vitro cultured cells will divide a finite number of times; cancerous cells can divide indefinitely.
Related topics	Mid-life and menopause in humans (D7) Aging theories (D9)

Aging and senescence

Lifespans, even of closely related species (e.g. mammals), can be very varied: for example a hamster 3 years, a human 80 years. Changes associated with age are readily observable: graying hair, wrinkles, presbyopia, declining muscle strength, etc. in humans. The science of **gerontology** attempts to study aging processes.

Strictly speaking, **aging** means any change associated with getting older, not necessarily deleterious (e.g. puberty). **Senescence** is a functional decline near the end of a lifespan. Most people use 'aging' as a synonym for senescence, implying deleterious change: an increased chance that an individual will die with time, or a decrease in an individual's ability to withstand environmental insults. Aging could be seen as a loss of vitality, the ability to sustain life.

Age-related changes are cumulative: death is sudden, but senescence involves a progressive increase in the probability of dying. Senescence is a fundamental, intrinsic property of all animals, and each species has a characteristic aging profile.

Survivorship curves in populations

Time of death in an individual tells us little about the process of aging, but times of death in a population are instructive: the chances of death occurring in a given time will rise if aging is taking place. This gives a mortality distribution.

(1) Assume a hypothetical **population born mature with no aging**. Assume 20% die each year from accidents or predation. A survival curve of the number of survivors against years gives an **exponential, hyperbolic curve** (*Fig. 1a*). This would be seen in a population where individuals all die of predation or accidents before they age.

(2) **Population which ages**. Here the curve is **rectangular** (e.g. a human survival curve; *Fig. 1b*) and is characteristic of a population where individuals age.

(3) **Wild populations**: here an **intermediate-type curve** is often seen, as effects of aging and predation/accidents take their toll.

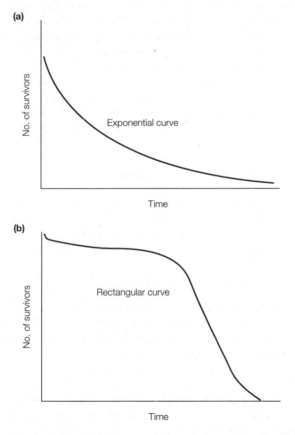

Fig. 1. (a) Population survivorship curve for a mature population with no aging (an exponential, hyperbolic curve). (b) Population survivorship curve for a population which ages (a rectangular curve).

Death distributions can also be shown by plotting numbers of deaths against time; q_x **(the age-specific death-rate)** can be calculated by dividing the number of individuals dying during each age interval by the number alive at the start of a given interval. In humans, q_x rises increasingly after 70 years: it could be said to be a measure of the chances of dying at a given age. However, population studies cannot predict how aging will occur in a given individual.

Aging in individuals

Different anatomical and physiological indices can be measured and trends with aging demonstrated, although some indices (e.g. gray hair in humans) may not be directly attributable to aging and be extremely variable between individuals. Test batteries of *several* different indices are best measured. There does seem to be an increasing imprecision of homeostatic controls with aging: for example blood glucose levels take longer to be restored to a set-point following a carbo-hydrate-rich meal. In humans, functions which show most change are those involving the co-ordinated activity of a number of organ systems: maximum breathing capacity falls by 50% between 20 and 90 years of age, whereas nerve conduction velocity falls by only 10%.

Population studies

Point-in-time studies measure aging indices in individuals of different ages at a given time (e.g. human 30 year olds, 50 year olds, 70 year olds in 1997); the danger is that one can assume that changes in average values from one generation to the next represent what occurs in individuals. Samples tend to be self-prejudiced for aging in that older individuals are 'fitter' survivors: if a factor is measured whose deterioration leads to an increased chance of dying, samples of older individuals are composed of those in whom the factor has changed least. Environmental changes can affect survival; for example, older humans were subjected to harsher conditions in terms of hygiene, disease exposure, etc. when they were young.

Longitudinal studies measure individuals from birth to death, but logistical problems are considerable: a longitudinal study of a population of humans born in 1897 would not yet be complete.

It should be noted that aging can, paradoxically, enhance survival: increased experience leads to more wisdom in avoiding hazards, immunological memory is enhanced, and decline in reproductive ability (e.g. after menopause in women, removing the dangers of childbirth) can increase vitality.

In vivo studies

Tissues can be divided into three types.

- **Renewable tissues**: for example those lining the gut. These are constantly being replaced: the precursors of such cells are stem cells. During aging, the generation time for such cells becomes longer (presumably giving more opportunities for functional defects to occur in the cells).
- **Expanding tissues**: for example liver cells. These do not normally divide but will so do if tissue damage occurs. Liver lobes can be regenerated quickly (within weeks in humans); older people regenerate their livers more slowly, cell division is less synchronous and the cells tend to have more chromosomal abnormalities.
- **Nonrenewable tissues**: for example brain cells. As aging occurs the number of neurons in the brain declines and the cells are not replaced.

Young tissue can be transplanted into older animals (and vice versa). The tissue continues to grow and the cells divide: successive transplants of the same tissue can result in a tissue far outliving the donor; there is some debate as to whether older animals are a more favorable environment for cells to flourish in than younger animals.

***In vitro* studies**   Cultured cells will often divide a finite number of times. Lung fibroblasts divide about 50 times, whereafter the generation time between division increases, cell debris accumulates and eventually division ceases (*Fig. 2*). This cessation is known as the **Hayflick limit** (after its discoverer) and it seems to be an intrinsic property of the cell clone itself. If old and young populations of cells are mixed, the old cells die while the young cells continue to divide, and adult lung cells enter a decline after fewer divisions than do fetal cells. Antioxidants (e.g. vitamin E, tocopherol) can increase the number of cell divisions, and it is significant that cancerous cell lines do not reach a Hayflick limit. (*In vivo* cells probably divide asynchronously, so that some earlier generation cells are always resting: thus gut cells can be renewed every day for 100 years in a human centenarian, clearly allowing far more time than might be predicted from the 50 or so synchronous divisions observed from *in vitro* studies.)

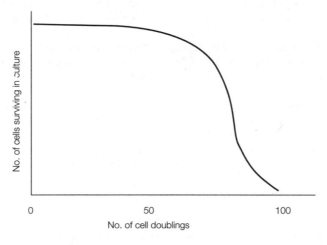

Fig. 2.   Common pattern of survival of human cells (e.g. fetal lung fibroblasts) in vitro.

D9 THEORIES OF AGING

Key Notes

Nature versus nurture	Theories of aging can be divided into those which are genetic (i.e. aging is encoded in the genome) and those which invoke chance or environmental causes.
Genetic theories	Evidence for genetic theories includes human longevity in different societies and eras, survival of gnotobiotic mice, parent/offspring and twin studies, progeria conditions and Hayflick limits in cultured cell lines.
Environmental theories	Evidence for environmental factors includes the shift of survivorship curves to the left in irradiated flies, enhanced longevity in diet-restricted rats and reduced longevity in overcrowded animals.
Specific theories of aging	Specific theories of aging can include some which are primarily genetic, and others which are enviromental, including somatic mutation of the genome, autoimmune decadence, molecular cross-linking of collagen, etc., free radical damage to membranes, faulty nucleic acid/protein synthetic machinery leading to error catastrophe, and faulty DNA repair facility in cells.
Overview of theories	No single theory can explain the overall aging process.
Evolution of senescence	Senescence probably evolved through weak selection against deleterious genes appearing late in life. An optimal balance between investment in reproductive capacity and repair involves levels of body maintenance that result in aging.

Related topics	Hypersensitivities, autoimmunity and immunization (C14)	Mid-life and menopause in humans (D7)
		Aging (D8)

Nature versus nurture

Theories of aging can be divided into those implying a genetic program and those implying that aging is environmentally determined: **nature versus nurture**.

Genetic theories of aging

Genetic theories note that as genes determine whether we are mice or humans, and mice live 2–4 years whereas humans live for more than 90 years, genes must play a role in determining lifespans. (There is a complex relationship which can be worked out between body size and brain size and average lifespan in different species, although there are exceptions, e.g. humans.) In favor of genetic theories for aging the following may be noted.

- Throughout history and in all societies, 70 years has been considered a respectable age and 80 years has been considered old (see Psalm 90, verse 10, reflecting realities of life in pastoral Palestine 3000 years ago!). In developed societies today, more people live into their 80s and 90s, but maximum lifespans have not increased. Centenarians are relatively rare: the oldest human age *reliably* recorded (in 1997) is 122 years.
- In most species, it is found that females on average live longer than males.
- 'Gnotobiotic' mice kept in a germ-free environment have a similar lifespan to 'normal' mice not shielded from dirt, pathogens, etc.
- Identical twins tend to have more similar lifespans than do nonidentical twins (although the former may have been brought up in more similar environments); studies on the lifespans of children of long-lived parents are equivocal (and such offspring may have been subjected to similar environmental stresses to those of their parents), and there is a little evidence to suggest that there is correlation between parent/offspring lifespans.
- Precocious aging, **progeria**, as in Hutchinson–Gilford's syndrome (where the child 'ages' from about 2 years old) and Werner's syndrome (where aging occurs in early childhood), does seem to be linked to autosomal recessive genes. However, it may be noted that such 'aging' differs from 'true' aging.
- Normal cells seem to have a method of 'counting' the number of cell divisions which they can undergo before they reach a Hayflick limit (see *Fig. 2* in Topic D8).

Environmental theories of aging

Environmental theories postulate that chance changes in molecules or organelles, or environmental insults, influence the course of aging. It has been noted that overcrowding seems to hasten the aging processes in many species of rodents, while dietary restriction in young rats delays maturity, the onset of age-related diseases and death. Lack of food in youth may slow down a developmental clock and prevent obesity, which certainly hastens aging and death.

Irradiation of fruit flies results in shifting of their survivorship curve to the left (*Fig. 1*), suggesting that such irradiated flies senesce early.

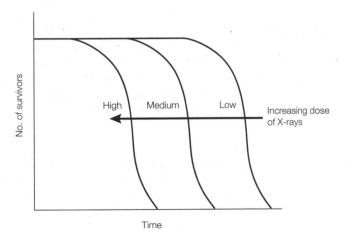

Fig. 1. The effects of X-irradiation on survival of fruit flies.

Specific theories of aging

Many **specific theories of aging** are postulated. Some involve genetic, programed components, others chance or environmental factors, yet others mixtures of several contributory components. Most are not mutually exclusive.

Somatic mutation

Environmental insults such as ionizing radiation or chemical mutagens cause mutations in the cells' genomes, leading to impaired cellular (and by a knock-on effect) organism function. X-irradiation of *Drosophila* does shift the survivorship curve to the left (*Fig. 1*); on the other hand, triploid frogs with one and a half times the normal amount of genetic material do not seem to age more quickly (or slowly), and persons exposed to high levels of ionizing radiation do not seem to age prematurely (although they do develop certain cancers).

Autoimmunity

The thymus involutes at puberty and the production of new T lymphocytes is impaired. T lymphocytes are responsible for many of the control and effector facets of immunity, and so there is a danger that the immune system becomes unbalanced and less capable of coping with infection. The self-censoring abilities of the immune system tend to break down in old age, leading to autoimmune conditions. This could lead to a state of 'autoimmune decadence' whereby the body's immune system starts to destroy the body (10% of 20-year-old humans and 88% of 80-year-olds have antinuclear autoantibodies). Autoimmunity does not explain aging processes in invertebrates which lack sophisticated adaptive immune systems.

Molecular cross-linking

Cross-linking between important skeletal molecules, observed during aging in collagen and elastin molecules, could impair biochemical and physiological cell efficiency.

Free radicals

These are chemical species with unpaired electrons in their outer orbitals and are transient intermediates in biochemical reactions; they can cause much damage to cell and organelle membranes. Many of the changes caused by free radicals are found in aging tissues; free radical scavengers such as vitamin E or hydrocortisone prolong the lives of cultured cells.

Catastrophe theories

There are several related theories. In essence, they note that DNA codes for messenger RNA which codes for proteins. Nucleic acid polymerases are essential enzymes to facilitate synthesis/transcription of new DNA or RNA. If 'errors' occur in polymerase transcription or translation of polymerases this will result in erroneous DNA and RNA, leading to further errors by a process of positive feedback, until a 'catastrophe' occurs. The existence of 'error-prone DNA polymerases' has been postulated in aging tissues, and aging cells do seem to have less specific enzymes and other defects in protein synthetic machinery, but older cells seem to support viral replication perfectly well.

Repair facilities

Many aging changes are ameliorated by repair processes, and aging cells seem less able to repair damaged DNA. Experiments show that the ability to repair DNA in cultured cells is related to the lifespan of the source species: humans are better repairers than are mice. In the disease xeroderma pigmentosum, ultra-violet radiation leads to fatal skin melanomas and other changes mimicking aging: this is related to a faulty DNA repair facility within the patients' cells.

Codon restriction

This theory (put forward by Strehler) postulates a programed decline in the repertoire of transfer RNA types within the cell during aging, progressively impairing translation at the ribosomes.

Overview of aging theories

It is probably true to say that there is no single cause of aging and no single theory will fit all the facts. A gradual breakdown in homeostatic control (the result of chance and environmental factors such as somatic mutation, free radical attack, autoimmunity, and errors in protein and nucleic acid synthesis played out against a genetic background which may include elements of a 'program') may be the key factor.

Evolution of senescence

There are a number of theories. The selection against genes which have a dele-terious effect late in life is weak: individuals will have already passed the genes to their offspring. Such genes tend to accumulate and are not selected out: they express themselves, together, later in life. This phenomenon we see as aging, exacerbated by an inability to resist environmental damage. (It is suggested that there may be a selective advantage in having a post-reproductive population of women in society to look after children while younger women are producing more children: this could explain the evolution of the menopause in women.)

Kirkwood and Holliday postulated a '**disposable soma**' theory in 1986: an individual's energy must be divided between maintenance and reproduction. Too much maintenance of the body will avoid aging but results in too low a reproductive rate; too little maintenance leads to rapid aging and the individual will not survive long enough to produce many offspring. A balance between maintenance and reproduction must be struck. When the complex mathematics of the theory are worked through, the optimum balance allows the maximum number of offspring for a level of maintenance which delays but does not prevent senescence.

FURTHER READING

There are many comprehensive textbooks of biochemistry and molecular biology, and no one book that can satisfy all needs. Different readers subjectively prefer different textbooks and hence we do not feel it would be particularly helpful to recommend one book over another. Rather we have listed some of the leading books which we know from experience have served their student readers well.

Sections A and B *General zoology*
Dorit, R.L., Walker, W.F. and Barnes, R.D. (1991) *Zoology*. Saunders College Publishing, Philadelphia, PA.
Hickman, C.P. and Roberts, L.S. (1990) *Biology of Animals*. Wm. C. Brown, Dubuque, IA.
Margulis, L. and Schwartz, K.V. (1988) *Five Kingdoms*. W.H. Freeman, New York.
Miller, S.A. and Harley, J.P. (1994) *Zoology*. Wm. C. Brown, Dubuque, IA.

Invertebrate zoology
Barnes, R.D. (1994) *Invertebrate Zoology*. Saunders College Publishing, Philadelphia, PA.
Brusca, R.C. and Brusca, G.J. (1990) *Invertebrates*. Sinauer, Sunderland, MA.
Meglitsch, P.A. and Schram, F.R. (1991) *Invertebrate Zoology*. Oxford University Press, Oxford.
Pechenik, J.A. (1996) *Biology of the Invertebrates*. Wm. C. Brown, Dubuque, IA.

Vertebrate zoology
Pough, F.H., Heiser, J.B. and McFarland, W.N. (1989) *Vertebrate Life*. Collier Macmillan, London; Macmillan, New York.
Romer, A.S. and Parsons, T.S. (1986) *The Vertebrate Body*. Saunders College Publishing, Philadelphia, PA.
Young, J.Z. (1981) *The Life of Vertebrates*. Oxford University Press, Oxford.

Section C *Comparative physiology*
Prosser, C.L. (ed.) (1991) *Comparative Animal Physiology: Volume I, Environmental and Metabolic Physiology*. Wiley-Liss, New York.
Prosser, C.L. (ed.) (1991) *Comparative Animal Physiology: Volume II, Neural and Integrative Animal Physiology*. Wiley-Liss, New York.
Randall D., Burggren, W. and French, K. (1997) *Animal Physiology: Mechanisms and Adaptations*. W.H. Freeman, New York.
Schmidt-Nielsen, K. (1997) *Animal Physiology: Adaptation and Environment*. Cambridge University Press, Cambridge.
(See also zoology texts listed under Sections A and B)

Human physiology
Clancy, J. and McVicar, A.J. (1995) *Physiology and Anatomy: a Homeostatic Approach*. Edward Arnold, London.
Vander, A.J., Sherman, J.H. and Luciano, D.S. (1994) *Human Physiology: the Mechanisms of Body Function*. McGraw-Hill, New York.

Immunology

Benjamini, E., Sunshine, G. and Leskowitz, S. (1996) *Immunology: a Short Course.* Wiley-Liss, New York.

Staines, N., Brostoff, J. and James, K. (1993) *Introducing Immunology.* Mosby, St. Louis, MO.

Turner, R.J. (ed.) (1994) *Immunology: a Comparative Approach.* John Wiley, Chichester.

Section D   Johnson, M. and Everitt, B. (1995) *Essential Reproduction.* Blackwell Science, Oxford.

Brookbank, J.W. (1990) *The Biology of Aging.* Harper & Row, New York.

(See also zoology and physiology texts listed under Sections A, B and C)

INDEX